New Trends in High Voltage Engineering
http://dx.doi.org/10.5772/intechopen.74085
Edited by Reza Shariatinasab

Contributors

Jin Li, Boxue Du, Zhaohao Hou, Dengming Xiao, Sushanta Paul, Reza Shariatinasab

Notice

Statements and opinions expressed in the chapters are these of the individual contributors and not necessarily those of the editors or publisher. No responsibility is accepted for the accuracy of information contained in the published chapters. The publisher assumes no responsibility for any damage or injury to persons or property arising out of the use of any materials, instructions, methods or ideas contained in the book.

First published in London, United Kingdom, 2018 by IntechOpen
IntechOpen is the global imprint of INTECHOPEN LIMITED, registered in England and Wales, registration number: 11086078, The Shard, 25th floor, 32 London Bridge Street
London, SE19SG – United Kingdom
Printed in Croatia

British Library Cataloguing-in-Publication Data
A catalogue record for this book is available from the British Library

Additional hard copies can be obtained from orders@intechopen.com

New Trends in High Voltage Engineering, Edited by Reza Shariatinasab
p. cm.
Print ISBN 978-1-78984-902-8
Online ISBN 978-1-78984-903-5

NEW TRENDS IN HIGH VOLTAGE ENGINEERING

Edited by **Reza Shariatinasab**

We are IntechOpen,
the world's leading publisher of
Open Access books
Built by scientists, for scientists

3,900+
Open access books available

116,000+
International authors and editors

120M+
Downloads

Our authors are among the

151
Countries delivered to

Top 1%
most cited scientists

12.2%
Contributors from top 500 universities

Interested in publishing with us?
Contact book.department@intechopen.com

Numbers displayed above are based on latest data collected.
For more information visit www.intechopen.com

Meet the editor

Reza Shariatinasab received a BS degree in Electrical Engineering from Ferdowsi University, Mashhad, Iran, in 2000, and MS and PhD degrees in Electrical Engineering from Amirkabir University of Technology, Tehran, Iran, in 2003 and 2009, respectively. In September 2007, he was a visiting scientist at Doshisha University, Kyoto, Japan. In 2009, he joined the Electrical and Computer Engineering Department, University of Birjand, Birjand, Iran, where he is currently an Associate Professor of electrical engineering. He is the author or coauthor of 3 books, and more than 110 scientific papers published in reviewed journals and presented at international conferences. His research interests include high-voltage engineering and lightning transients.

Contents

Chapter 1

Introductory Chapter: New Challenges in High-Voltage Engineering

Reza Shariatinasab

Additional information is available at the end of the chapter

http://dx.doi.org/10.5772/intechopen.80623

1. Introduction

Since the advent of electricity industry, a feasible way to overcome power losses and voltage drop in transmission lines has been to increase line voltage level of the transmission network. A higher voltage leads to a higher efficiency and less loss as well as more transmission capacity of the line and extends the value of transmitted power over longer distances. High-voltage engineering is the science of planning, operating, and testing high-voltage electrical devices and designing the insulation coordination in order to ensure the reliable operation of the power network. Therefore, high-voltage engineering provides the access to electrical energy for consumers far away from power generation units. This branch of science develops and optimizes operating characteristics of internal and external insulators.

The appearance of semiconductor valves and attractive aspects of direct current network have been led to the development of high-voltage DC transmission lines (HVDC). Some advantages of HVDC transmission systems are their high dynamic stability, ability to be connected to large DC renewable sources and DC micro-grids, etc. Thus, developing large load centers and restricting high-voltage installation places makes it possible to extend HVDC underground cable lines. This attracts the attention of high-voltage engineers to improve the characteristics of cable insulators and other DC equipment's insulation.

As the concerns about global warming, greenhouse gas emission, and recycling the artificial wastes have been growing recently, different industries are forced to produce recyclable devices with lower environmental consequences. High-voltage researchers have extensively investigated nanopolymer cable insulators and environment-friendly materials to replace existing insulation materials. Cross-linked polyethylene (XLPE) is a well-known high and ultrahigh voltage cable insulation with interesting features such as excellent electrical and mechanical characteristics, high reliability, and low cost. However, because of recycling difficulties and environmental pollution, high-voltage scientists are trying to replace XLPE by nanopolymers like polypropylene/inorganic nanocomposites. The nanocomposites have better thermal and electrical characteristics than XLPE. Of course, nanopolymer manufacturing technologies are progressing, and there are many challenges in this way with regard to interactions between temperature, electric field, space charge, and DC volume resistivity under multifield coupling. It is necessary to review polypropylene and polypropylene/inorganic nanocomposites, as well as the opportunities and challenges for using them in HVDC cable insulation.

Another development in high-voltage engineering for environmental compatibility is realized through replacing insulation gas SF_6 by environmentally friendly gas CF_3I in gas-insulated stations (GISs) and high-voltage circuit breakers. SF_6 is an inert and electronegative gas with an excellent insulation strength. When a high-voltage circuit breaker opens short current from the source, the contacts are separated, while SF_6 fills contact gaps; the gas experiences severe arc discharge for extinguishing the electrical arc. After extinguishing, because of an extremely high temperature, a plasma channel is formed. The powerful gas attracts plasma energy to avoid circuit breaker destruction. However, for avoiding global warming due to greenhouse gases, CF_3I is used, as an environmentally friendly gas. CF_3I is a colorless and volatile gas, and because of high boiling point of pure CF_3I liquid, it cannot be used as a dielectric substance. Therefore, CF_3I mixed with CO_2 or N_2 offers a dielectric strength by 75–80% higher than that of SF_6. Thus, extended researches have been performed to replace SF_6 by CF_3I (mixed with CO_2 or N_2) with different ratios, considering operating characteristics.

Because the conductors of a high-voltage cable line are packed in a confined space, the resultant voltage gradient causes an intense electric field, which in turn leads to high capacitance. Therefore, the cable current flow contains significant capacitance component rather than inductive component; this results in higher voltage at downstream node. Due to equivalent capacitance and inductance, natural frequency of a high voltage cable can be excited by harmonic content of a transient impulse that leads to destructive overvoltage known as "resonance" phenomenon. Actually, when natural frequency of equivalent capacitance and inductance is equal to one of harmonic components of transient switching impulse waveform, the resonance occurs. Depending on equivalent circuit in the branch where resonance occurs, resonances are categorized into three groups, i.e., series, parallel, and compound. The series resonance leads to voltage drop across the resonating branch. A very large impedance appears across the parallel branch that can create destructive overvoltage by a flowing small leakage current. Therefore, designers must consider the resonance in the planning of the insulation coordination of the networks.

One of the main challenges in designing and operating high and ultrahigh voltage transmission lines is the occurrence of short circuit that makes bulk transmitting lines out of service and can destabilize the overall network. The overvoltage caused by lightning or switching is one of the main reasons for short-circuit occurrence along transmission lines. Therefore, planning engineers perform appropriate insulation coordination for line insulators against this kind of overvoltages with considering voltage level and line characteristics. Nevertheless, pollution aggregation on the insulator surface decreases the electric withstand. The pollution can be formed by combined dust, humidity, and salt in industrial or coastal areas where maintenance intervals may be very long. Therefore, planning engineers must know environmental circumstances in addition to electrical characteristics of transmission lines. Of course, this problem can be solved by increasing insulation strength, using more compatible insulators and washing the line insulators periodically.

Artificial tests of manufactured products are necessary in industrial applications as well as scientific studies. High-voltage devices need testing in manufacturing process, before sale, laboratory, and quality control stages, after installation for ensuring safe transportation/installation and predetermined maintenances. High-voltage testing is performed under loading or artificial impulses such as switching or lightning waveforms. The loading tests are performed in order to detect capacitance discharge of dielectric or partial discharge and insulation resistance. The impulse tests are performed to determine insulation strength of an external insulator that is not considered sufficient. The external insulator is designed for outdoor usages at different distances from the ground or on top of another isolator as in porcelain cover of bushings, buses, and sectionalizing insulators. Therefore, testing voltage is applied to a device more than one time, and then, complete flashover probability is calculated. The insulation strength has a stochastic nature, so it can be evaluated by statistical approaches. Characteristics of testing waveforms must be determined in a way that actual conditions can be realized.

Author details

Reza Shariatinasab

Address all correspondence to: shariatinasab@birjand.ac.ir

Electrical and Computer Engineering Department, University of Birjand, Birjand, Iran

A Review of Polypropylene and Polypropylene/ Inorganic Nanocomposites for HVDC Cable Insulation

Boxue Du, Zhaohao Hou and Jin Li

Additional information is available at the end of the chapter

http://dx.doi.org/10.5772/intechopen.80039

Abstract

Due to its excellent electrical and thermal performance, as well as satisfying the needs for developing the environmentally friendly and recyclable cable insulation material, polypropylene has caused widespread concern. Nanodoping can effectively improve the electrical, thermal and mechanical properties of polypropylene nanocomposites, which provides a new method to solve the problems in its application in HVDC cable insulation. This chapter introduces research achievements on polypropylene and polypropylene/inorganic nanocomposites, which states the effects of nanodoping on the electrical properties, such as space charge behaviors, electrical tree aging, breakdown strength, etc. thermal conductivity and mechanical properties of the polypropylene and its multi-blends. The aging mechanism under different conditions is also discussed. The analysis shows that the surface treatment of nanoparticles can reduce the aggregation of nanoparticles and strengthen the interface effect, thus improving the comprehensive properties of polypropylene nanocomposites. This chapter also summarized the feasibility and future development of the polypropylene and its nanocomposites application in the insulation of HVDC cables.

Keywords: polypropylene, nanocomposites, HVDC, cable insulation

1. Introduction

With the constant construction and commissioning of HVDC transmission projects worldwide, the problems exposed by the production and operation of high-voltage DC cables using XLPE as insulating materials are also increasing. Crosslinking agents used in the production of XLPE cables and by-products from the cross-linking process may be introduced into the insulation layer, making the space charge accumulation under the DC electric field more serious, thereby accelerating the insulation aging. In addition, the cross-linking process used in the manufacture of XLPE cables is inherently inefficient and inefficient. Moreover, after the XLPE cable has reached the end of its useful life, it is very difficult to recycle and reuse the insulating waste. Incineration processing not only pollutes the environment, but also wastes resources [1].

The research of domestic and foreign researchers on the large-capacity environment-friendly DC cable insulation material mainly focuses on polyethylene (PE) and polypropylene (PP)-based materials. Compared with PE, PP has a relatively high melting point, which can meet the demand of cables operating at higher temperatures, and has higher breakdown strength and volume resistivity, which is of great significance for increasing the operating voltage level of the cable and the line ampacity. However, polypropylene materials have strong brittleness and rigidity, poor resistance to low temperature impact, and low thermal conductivity. The operating conditions of high-voltage DC cables are complex, and the insulation medium is affected by the strong electric field with constant polarity, the temperature field generated by conductor heating, and the mechanical stress generated externally or internally in the medium. Therefore, the research on polypropylene-based environment-friendly insulating materials needs to be conducted to meet electrical, thermal and mechanical performance requirements [2].

In recent years, it has been found that nanoparticles have excellent performance in improving the properties of polymer materials because of their quantum size effect and large specific surface area [3]. Since 1994, when Lewis proposed the concept of nanodielectrics [4], scholars from various countries have extensively studied the improvement of the properties of polymer insulating materials after adding nanoparticles and their improvement mechanisms. The thermal performance, as well as the mechanical properties such as tensile strength and elongation at break, are not exactly the same. Most scholars believe that the nanoscale transition region between the polymer and the nanofiller, i.e. the interface, is a key factor affecting the performance of the nanocomposite [5–7]. The characteristics of the polymer matrix and the characteristics of the nanofiller determine the interface structure and properties of the composite material. Although many scholars have proposed different models to explain this, there is still no conclusion.

This chapter summarizes the research results at home and abroad, introduces the feasibility of polypropylene and its application in high-voltage DC cables, and discusses the role and mechanism of nanofillers in improving the electrical, thermal and mechanical properties of polypropylene monomers and multi-component blends. The effect of aging conditions on the properties of polypropylene nanocomposites was summarized, and the research on polypropylene-based nanocomposites for high voltage DC cables was summarized and forecasted.

2. Polypropylene and its feasibility study for insulation materials of high-voltage DC cables

2.1. Physical and chemical properties of polypropylene materials

Polypropylene is a thermoplastic resin obtained by polymerization of propylene as a monomer, which has a regular structure, high crystallinity, good corrosion resistance, and excellent heat resistance. Polypropylene can be divided into isotactic polypropylene (iPP), syndiotactic polypropylene (sPP) and atactic polypropylene (aPP) according to its methyl group position. The molecular structure of the three polypropylenes is shown in **Figure 1**.

The melting point of polypropylene can reach more than 150°C (different melting point of different brands), about 40–50% higher than that of polyethylene, and the long-term working temperature can reach 90°C. Polypropylene is a non-polar material with high breakdown strength (mostly around 300 kV/mm), high bulk resistivity (mostly around 10^{16} Ω m) and insignificant change with temperature, can be in the same insulation. Polypropylene has less space charge accumulation and the charge injection has a higher threshold electric field. Polypropylene hardly absorbs water, so its insulation properties are less affected by the ambient humidity.

In addition, polypropylene materials have high mechanical strength without cross-linking treatment, and are typical thermoplastic materials that can be recycled and used in line with the development needs of environmentally friendly cable insulation. However, the polypropylene material itself also has some disadvantages, such as large low-temperature brittleness, poor aging resistance, low thermal conductivity, etc., which have certain limitations on the application of DC cable insulation.

2.2. Feasibility of polypropylene applied to insulation material for high-voltage DC cable

Polypropylene has excellent dielectric and heat resistance properties. As early as 2002, some scholars have studied the feasibility of applying it to the main insulation material of power cables. Among them, the Japanese researcher Kurahashi found that sPP-based insulation, adding PE and antioxidant blends made of 0.6 and 22 kV cable, at different temperatures in the cable line AC breakdown strength and dielectric loss can meet the requirements of practical applications [8]. Yoshino et al. found that the electrical, thermal, and mechanical properties of sPP were superior to those of iPP, aPP, and PE materials [9]. The 22 kV cable prepared by the authors using sPP and elastomer blends has excellent electrical properties and shown a high

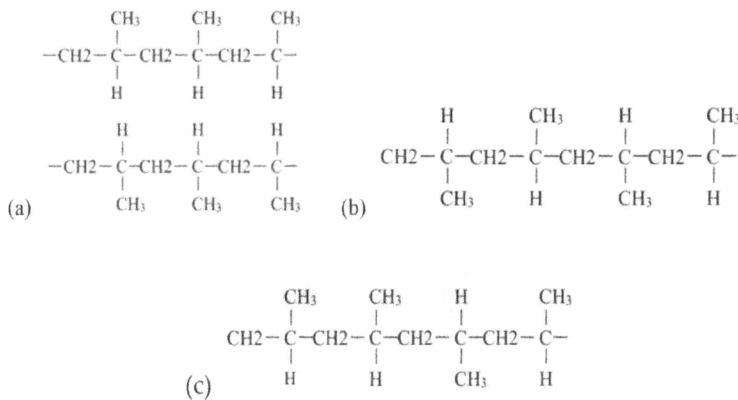

Figure 1. Molecular structure of the three kinds of polypropylene: (a) iPP, (b) sPP and (c) aPP. (a) (b) (c).

impulse breakdown strength than XLPE at 25, 90, 110°C, as shown in **Figure 2**. Hosier et al. found that insulating materials obtained by blending iPP with ethylene-propylene copolymers exhibit good mechanical toughness and electrical properties [10]. As shown in **Figure 3**, the PP (as labeled with H) shows a lower temperature dependence- conductive current than that of PE (as labeled with 1,2,3,4,5) [11].

At present, the commercial application of polypropylene-based materials as the main insulating material for high-voltage DC cables is still in the research and development stage. In 2010, Belli et al. of Prysmian, Italy disclosed high-performance thermoplastic elastomer insulation material HPTE (High Performance Thermoplastic Elastomer) [12] developed based on polypropylene material. The study found that P-Laser cable developed based on HPTE material is more traditional than traditional. XLPE cables have better electrical properties and have better mechanical properties than polypropylene.

Polypropylene materials have excellent comprehensive performance, and the research in the field of main insulation of high-voltage DC cables also shows great potential, but there is still a certain distance from practical application. Polypropylene materials have insufficient flexibility and poor low temperature toughness at room temperature and cannot be directly used for main insulation of cables. Moreover, most researchers are concerned about the improvement of the mechanical properties of polypropylene. The study of the dielectric properties of polypropylene based materials is not comprehensive enough, and space charge, electrical branch, etc. are not considered.

In the operation of high-voltage DC cables, there are problems such as aging and breakdown of electrical branches due to the space charge accumulation and internal electric field distortion problems [13]; the electric field reversal that may be caused by the influence of heat

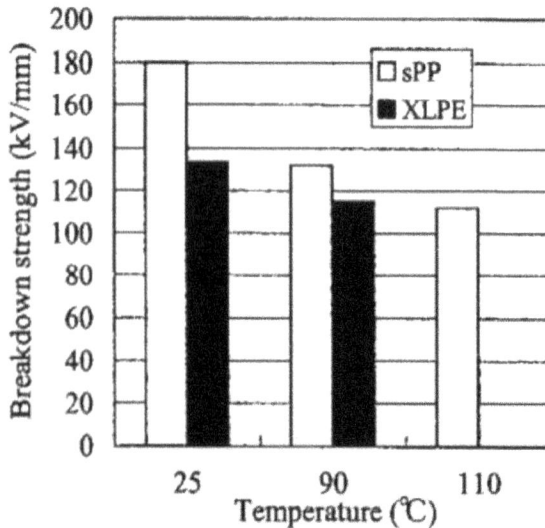

Figure 2. Impulse breakdown strength of 22 kV class sPP and XLPE cable.

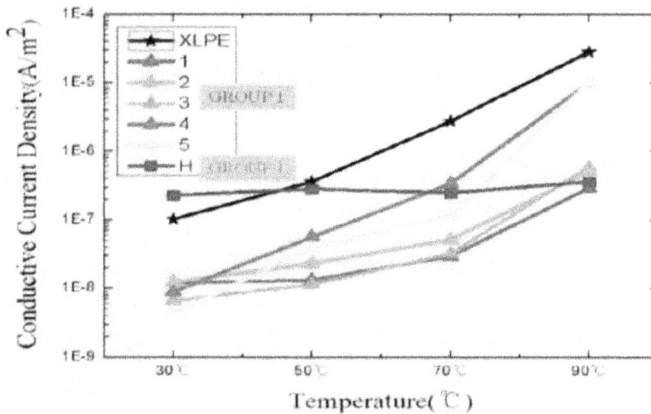

Figure 3. Temperature dependence of conductive current.

dissipation and temperature gradient of the insulation layer, Problems such as the degradation of electrical performance and service life; and the problems of internal defects caused by mechanical stress, etc. If polypropylene is used as insulation material for high-voltage DC cables, it must be modified to improve the above-mentioned deficiencies in dielectric, thermal and mechanical properties. Nanoparticles modification can significantly improve the electrical, thermal and mechanical properties of solid dielectrics. Many scholars have conducted useful explorations on the performance of polypropylene-based nanocomposites.

3. Research status of polypropylene-based nanocomposites for HVDC cable insulation materials

3.1. Study on dielectric properties of polypropylene nanocomposites

3.1.1. Space charge

The space charge effect is a key issue in the development of high performance DC insulation materials [14]. When the high-voltage DC cable runs normally, the electric field with the same polarity and high strength acts on the insulating medium for a long time, resulting in the accumulation of the space charge of the insulating layer and the distortion of the internal electric field. The distorted electric field can cause partial discharge in the medium, accelerate the aging of the polymer material and the growth of the electric branch, and eventually lead to insulation breakdown failure, which seriously affects the performance and service life of the cable. Since the mid-1990s, scholars such as Suzuoki and others have found that the pre-applied voltage causes the distortion of the internal electric field due to the space charge accumulation in polypropylene and reduces the dielectric breakdown strength [15]. Therefore, how to improve the space charge characteristics of polypropylene insulation materials is an important issue for the development of polypropylene-based insulation materials for high-voltage DC cables.

Most scholars believe that the interface region formed between the polymer matrix and the nanoparticle in the nanocomposite material introduces a large number of traps, changes the trap energy level of the composite material, and has an important influence on the space charge injection, migration and dissipation behavior. However, due to the complexity of the interface behavior (affecting polymer crystallization, changing the internal stress of the medium, etc.), and the ability to directly observe the microstructure and mechanism of action in the interface region, many scholars have proposed different models for this, such as proposed by Lewis et al. The nanodielectric "dielectric double layer" structure [5] (as shown in **Figure 4**), the multi-core model proposed by Tanaka [6] (as shown in **Figure 5**), Kindersberger et al. The volumetric model [7], to a certain extent, helps to speculate and explain the superior performance of nanocomposite dielectric materials, but it has not yet reached a conclusion.

Montanari et al. studied the charge trapping behavior of nanocomposites of iPP and sPP with synthetic Montmorillonite (MMT) nanoparticles. Compared with pure PP, the charge trapping ability of nanocomposites significantly increased. The reduction of the space charge accumulation under the electric field shows that the insulation performance of nanocomposites has been improved overall [16].

Due to the large surface energy of nanoparticles, nanoagglomeration is likely to occur during the preparation of nanocomposites, which not only reduces its dispersivity and interaction with the polymer matrix, but also aggravates the composite materials. Accumulation of space charge. Fuse et al. found that the introduction of ionic groups during the dispersion of nano-clay particles into polypropylene-based materials aggravated the space charge accumulation of the composite [17].

In order to solve the problem of nanoagglomeration, researchers have done a lot of research work and achieved certain results. It has been found that by adjusting the conditions for preparing the nanoparticle and the polymer matrix, such as temperature, the dispersion effect of the nanoparticle can be improved. Li et al. obtained different nanodoped iPP/MgO nanocomposites through mechanical blending at six different temperatures. It was found that when the temperature was 200° C., the nanoparticle dispersion was good, and MgO nanoparticles were in the composite. Nucleation occurs while suppressing the accumulation of space charge [18].

The use of coupling agents, surfactants, grafting, in-situ polymerization and other means of surface treatment of nanoparticles can reduce nanoaggregation and promote the dispersion of nanoparticles in the polymer and its interaction with the matrix. Abou-Dakka and others filled with synthetic nano-mica particles and natural montmorillonite nanoparticles modified

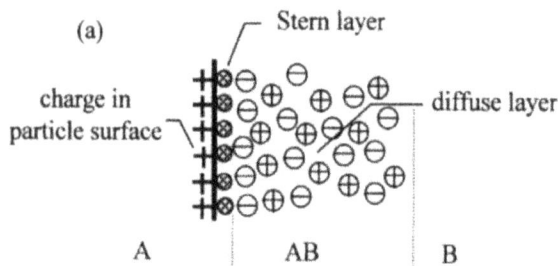

Figure 4. The diffuse electrical double layer produced by a charged particle A in a matrix B containing mobile ions.

The first layer

The second layer

The third layer

Nano-particle

Matrix

40 to 100 nm

Inter-particle Distance

(surface to surface)

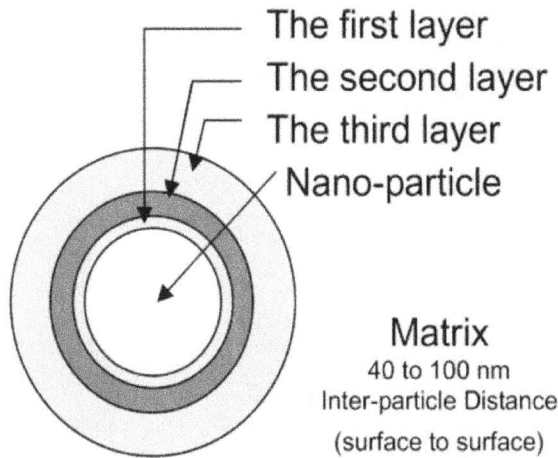

Figure 5. Multi-core model for nanoparticle - polymer interfaces.

with silane coupling agent make the trap-belt of polypropylene-based composite materials move to the shallow and the charge dissipation rate is much faster than that without filling. The charge in shallow traps at the unipolar level can be effectively suppressed in reverse polarity, and the charge in deep traps is also significantly constrained [19]. Zhou et al. found that a large number of shallow traps introduced into polypropylene composites by surface-modified TiO_2 nanoparticles replaced the original deep traps in PP, which in turn enhanced carrier migration and improved space charge accumulation [20].

3.1.2. Aging of electrical trees

During the production and operation of high-voltage plastic DC cables, defects such as impurities, voids, and molecular bond bonds may be generated in the insulating medium. When the polymer insulating material is placed under a high-intensity electric field for a long period of time, electric field concentration and partial discharge caused by defects or the like easily cause dendritic partial damage in the insulation, and the dendritic microchannels grow along the direction of the electric field to form electric branches. It can penetrate the entire insulation and cause breakdown accidents [21].

Du et al. found that dendritic electrical branches can be grown when PP is subjected to a pulse voltage of 12 kV and a frequency of 400 Hz at different temperatures. Compared to XLPE, the electrical branches in polypropylene are more difficult to produce and the growth speed and size are smaller [22], as shown in **Figure 6**, which is of great significance for improving the reliability of cable operation. Holto et al. observed that there were single and multiple branches growing before sPP breakdown [23]. Therefore, research on insulating materials suitable for high-voltage DC cables requires the suppression of their electrical branches.

In nanocomposites, when the dendrites grow to nanometer positions, it is generally difficult to pass through the nanoparticles, and the branch channels bypass the nanoparticles or stop growth, especially when the nanoparticles are lamellar structures. In addition, the added nanoparticles have a large specific surface area, and the tiny holes generated around

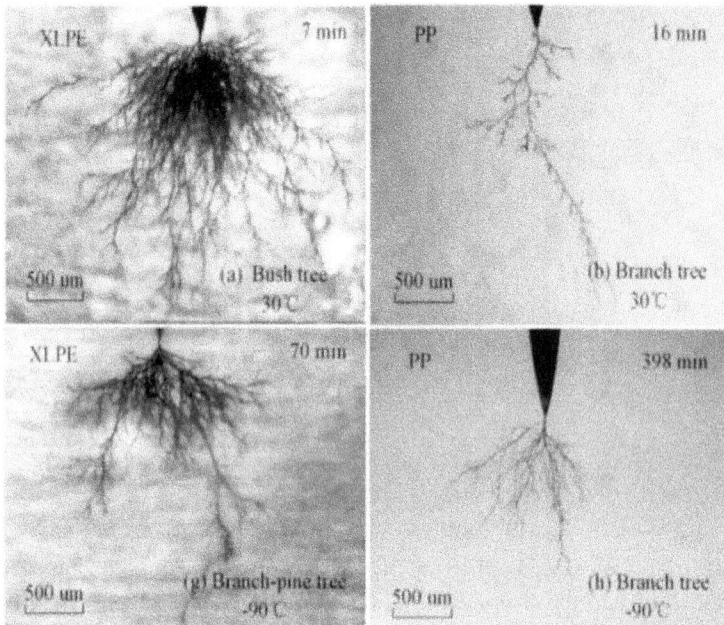

Figure 6. Electrical trees in XLPE and the PP.

the particles increase the number of branches of the electric tree, consume the energy of the development of the electric tree, and increase the probability of appearance of the jungle-like electric tree structure, delaying the electric branch. Growth rate and breakdown time.

3.1.3. Breakdown strength

The compressive strength of insulating materials is an important indicator for evaluating the electrical performance of cables. Nanodoping can increase the breakdown strength of polymers. It is of great significance for the development of polypropylene-based cable materials. According to the "dielectric double layer" structure model, filling a certain amount of nanoparticles can increase the trap energy level and trap density of the composite material, so that the same polarity charge accumulates on the surface of the material, the injection amount of carriers is reduced, and the material is weakened. Distortion of the electric field caused by the accumulation of charge in the space increases the field strength required for the composite to reach breakdown. At the same time, the nanoparticles can fill the spherulite gap formed during the crystallization of the polymer, blocking the transport and injection of charge.

At present, many literatures have studied the breakdown characteristics of polypropylene nanocomposites. Takala et al. found that POSS nanoparticles can be filled with polypropylene spherulite gaps to block the transport of charge and greatly increase the breakdown strength of nanocomposites [24]. Takala et al. also found that compared with pure PP, the AC/DC breakdown field strength of PP/SiO2 nanocomposites was significantly improved, and the DC breakdown strength increased by 52.3% [25].

When studying the breakdown properties of composites, the analysis of microstructures is also worthy of attention. Virtanen et al. found that the dispersion levels of nano-$CaCO_3$ particles with different doping levels were almost the same in the polypropylene matrix, while the density of particles caused by nanoagglomeration increased exponentially with the increase of nanodoping concentration. The dc breakdown strength of the composites is the largest at a mass fraction of 1.8%, and the density of the particles increases as the concentration of nanoparticles increases and decreases [26].

To improve the breakdown strength of polypropylene, SiO_2, Al_2O_3, $CaCO_3$, POSS and other nanomaterials are used, so that the amount of the composite material to achieve the best breakdown characteristics is also different. The use of nanosurface treatment to improve the agglomeration and its combination with the matrix The study of improving the breakdown strength of composites can be further explored.

3.2. Thermal conductivity of polypropylene nanocomposites

When the high-voltage DC cable runs normally, the heat of the cable conductor causes the temperature distribution of the insulating medium to decrease from the inner to the outer gradient, which has a great influence on the insulation layer charge transport, electric field distribution, and service life [27]. The electric field intensity distribution of the cable insulation layer under direct voltage is proportional to the resistivity, while the electric resistance of the polymer insulation material decreases with increasing temperature, so that the electric field strength outside the cable insulation layer is higher than the inside. At the same time, the effect of the temperature gradient aggravates the injection and migration of the cable conductor into the dielectric, further enhancing the electric field strength outside the insulating layer. When the accumulation of heteropolar space charge in the medium is serious, it can cause the reversal of the electric field, and even insulation breakdown. Moreover, the elevated operating temperature of the cable accelerates the aging of the polymer insulation and shortens the service life of the cable. Therefore, it is of great significance to improve the thermal conductivity of polypropylene insulation materials, to extend the service life of cables, and to increase the operating voltage and operating temperature of cables.

Filling metal or inorganic fillers with high thermal conductivity is an important method to improve the thermal conductivity of polymers. The thermal conductivity of composite materials is not only related to the intrinsic thermal conductivity of fillers and polymers, but also affected by the size, shape, and dispersion state of the filler. Generally, the smaller the size of the thermal conductive filler, the larger the interface area between the filler and the matrix, the more severe the phonon scattering, the higher the thermal resistance at the interface, and the worse the thermal conductivity. In terms of shape, the sheet-like or whisker-like heat-conductive filler has a large specific surface area, and it is easier to form a heat-conducting channel in the polymer matrix and improve the heat-transfer efficiency of the composite material.

For fillers with higher intrinsic thermal conductivity, such as nanosized boron nitride (BN), the thermal conductivity of the composite can be maintained at a high level even if the filler size is reduced to the nanoscale. Couderc et al. found that when the mass fraction of hexagonal boron nitride (hBN) particles is 50–80%, the thermal conductivity of composites is significantly higher than that of pure PP, and is almost unaffected by the concentration of boron nitride. At the same time, the introduction of nano-hBN particles can promote the dispersion of hBN micro-particles in the polypropylene micro-nano composite materials, thereby forming a more compact structure and weakening the thermal aging rate of the sample [28].

The introduction of nanosized thermally conductive fillers not only improves the thermal conductivity of the polypropylene material, but also improves the electrical properties of the composite material by the interfacial effect between the nanoparticles and the matrix. Du et al. found that the addition of nano-BN in polypropylene can significantly increase the thermal conductivity of composites. The direct-current body breakdown strength of PP/BN nanocomposites is not only higher than that of pure PP and PP/BN micron composites. As the content of nanomaterials increases, it increases accordingly [29].

Fukuyam et al. found that the thermal conductivity of grafted PP/SiO2 nanocomposites was higher than that of untreated composites [30]. The authors found that the thermal conductivity of the PP/SiO$_2$ nanocomposites treated with grafts was analyzed by using the three-phase model and considering the interface layer. The interface layer is the main channel for thermal conduction of the iPP-grafted molecular chains. **Figure 7** shows TEM images of nano-SiO$_2$ particles doped with grafted and ungrafted.

3.3. Study on mechanical properties of polypropylene nanocomposites

When the high-voltage DC plastic cable is working under load, the temperature distribution in the insulating layer is not uniform due to the conductor heating, and the thermo-mechanical

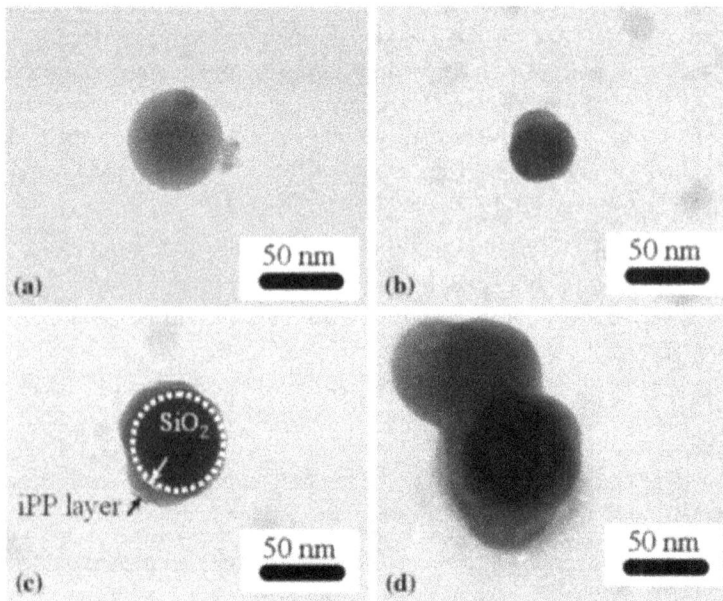

Figure 7. TEM images of n-SiO$_2$ (a) and g-SiO$_2$ (b–d) particles with various molecular weight of iPP.

performance changes due to thermal expansion and thermal stress can occur in the dielectric; the cable manufacturing and laying process may also cause Concentration of stress in insulating media. The role of mechanical stress can produce air gaps or micro cracks in the medium to form defects. The defects are easily discharged under the action of an electric field, causing serious breakdown of the material and threatening the operational safety of the cable. Polypropylene has high crystallinity, regularity, and excellent resistance to bending fatigue, but its high brittleness and low impact strength, especially in low temperature environments, need to be toughened and modified before it can be applied to the insulation of high-voltage DC cables.

Nano-doping can significantly increase the mechanical strength and fracture toughness of polypropylene materials, and reduce the generation of micro cracks and internal air gaps. The surface treatment or grafting process causes a physical adsorption or chemical reaction between the nanoparticles and the polymer matrix or the surface modifier, improving the mechanical properties of the polymer material. The grafted nanoparticles can promote the crystallization of the composite material, strengthen the link between the matrix and the nanoparticles, and improve the mechanical properties of the composite material. Umemori et al. found that the grafted PP-g-SiO2 nanoparticles have improved dispersibility in polypropylene and nucleation of PP crystals; the end of the grafted chain directly connects the PP matrix and SiO2 nanoparticles. By bridging together, the Young's modulus and tensile strength of composites can be greatly improved when the nano-mass fraction is 2.3% [31].

The introduction of nanoparticles can avoid the problem of reduced fluidity of the system caused by simple elastomer toughening, and can also promote the dispersion of elastomers, resulting in a synergistic toughening effect. Lee et al. found that the dispersion of nano-SiO2 particles treated with maleic anhydride in the matrix was more uniform; while the introduction of nano-SiO2 particles increased the shear force during the mixing process, and the particle size of the elastomer POE was further decomposed. Small, more uniform dispersion; synergistic toughening effect of nanoparticles and elastomers, so that the bending strength and Young's modulus of the composite material significantly improved [32].

3.4. Effect of nanofillers on properties of polypropylene composites

The brittleness of polypropylene is a key factor limiting its use as a high voltage DC cable insulation. Many studies have shown that the blending of polypropylene with elastomers is an effective means to improve the mechanical toughness. However, it was also found that the elastomer introduced more interfaces and traps in the blends, leading to more space charge accumulation, lower breakdown strength, and other insulation problems in the blends. Considering that nanoparticles have excellent performance in improving the insulation properties of polypropylene monomer, many scholars have studied the effect of nanoparticles on polypropylene multicomponent composites.

Du et al. found that POE can significantly improve the toughness of PP, but the space charge accumulation of PP/POE blends increases, and the breakdown field strength also decreases significantly; nano-ZnO particles maintain good mechanical toughness after PP/POE introduction. It also increased the tensile strength of PP/POE; filled nano-ZnO particles increased the trap level density of PP/POE, reduced the charge injection, and thus suppressed the space charge accumulation [33], as shown in **Figure 8**. Compared with PP/POE blends, PP/POE/ZnO nanocomposites have lower dielectric constants and higher breakdown and volume resistivity.

Zhou et al. mixed surface-modified nano-MgO particles into PP/POE. The study found that the DC breakdown strength and space charge suppression ability of PP/POE/MgO nanocomposites were enhanced, and the synergies between nanoparticles and POE were also increased. Tenacity results in a significant increase in the tensile strength and elongation at break of the composite [34]. Dang et al. found that the introduction of nano-ZnO reduced the number of deep traps in the PP/PER/ZnO nanocomposites, promoted the dissipation of heteropolar charges in the composites, and effectively inhibited the formation of space charge [35].

3.5. Effect of nanofillers on the properties of polypropylene composites under aging conditions

When polypropylene is applied to the insulation material of high-voltage DC cables, the problem of material aging is an important index for evaluating its insulation performance and service life. The combined action of the electric field and heat during normal operation of the cable accelerates the aging of polymer insulation performance. The entrapment and extinction of the electric charge in the insulating medium of the cable is accompanied by the release and transfer of energy. The ultraviolet radiation generated at this time can lead to polymer degradation. The conductor heat directly acts on the dielectric insulating layer, and the weak link of the covalent bond of the polymer molecular chain first initiates free radicals and undergoes chain reaction, resulting in chain scission and destroying the structure of the polymer. A large number of tertiary carbon atoms are distributed on the molecular chains of polypropylene polymerized from propylene. Under aerobic conditions, tertiary carbon atoms are extremely unstable and easily convert to very active tertiary carbon radicals, resulting in PP chain growth, chain degradation. In the process of developing polypropylene-based high-voltage DC cable insulation materials, the effect of nanofillers on their performance under electrical and thermal aging has attracted the attention of relevant scholars.

The nanometer lamellar structure has mechanical protection and physical barrier effect on the polymer, and the interface interaction between the nanoparticle and the polypropylene restricts the movement of the amorphous phase molecules. These enhance the commonness of

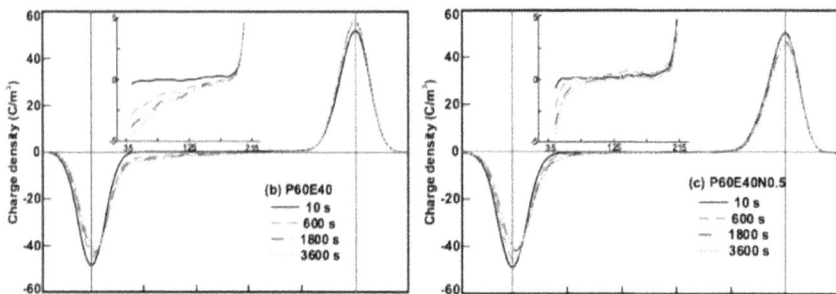

Figure 8. Space charge distribution of PP/POE(left) and PP/POE/Nano-ZnO(right).

the polypropylene nanocomposite in the electric field and temperature field. Under the effect of thermal aging resistance and insulation performance. Moreover, lamellae-structured nanoparticles have a higher surface energy, so that carriers are trapped by interface traps between the nanoparticle and the polymer matrix during the transition, limiting the migration of electrons and holes. In the composite material, when the nanometer doping amount is large, the overlap of the "media bilayer" structure between the nanoparticles can be enhanced, the space charge transfer is promoted, and the electrical conductivity of the nanocomposite material becomes large. Guastavino et al. found that the nanometer MMT mass fraction of polypropylene nanocomposites with the mass fraction of 10% was the highest in the mid-term (about 277 h) electrical aging test [36]. Bulinski et al. added polypropylene tetra-mica nanoparticles containing synthetic tetrasilicic fluormica nanoparticles to a DC −40 kV/mm electric field intensity at room temperature and 90°C, respectively, and accumulated aging for 500 h. The tensile strength is still about 12% higher than that of pure polypropylene, and there is no significant change in the dielectric loss factor, and the DC conductivity is only slightly increased [37].

4. Conclusions

The development of high-voltage direct current (HVDC) transmission technology places higher demands on the cable's current-carrying capacity, voltage operating level, and operating temperature. Polypropylene has good heat resistance, high melting point and long-term working temperature, excellent electrical properties such as breakdown strength, volume resistivity, space charge, etc. No cross-linking is required to simplify the production process, thermos plasticity can be recycled, and it is in line with large-capacity and environmental protection Technical requirements for DC cables. However, polypropylene has the disadvantages of poor low-temperature impact performance and low thermal conductivity, as well as the problem of space charge accumulation and aging of the polymer in the DC field. Therefore, polypropylene needs to be modified in order to meet the electrical, thermal, and mechanical properties of the cable insulation material under the complex working conditions of high-voltage DC. Nano-doping can effectively improve the overall performance of polypropylene monomer and multi-component blended composites, such as suppression of space charge accumulation, resistance to aging of electrical branches, improvement of dielectric strength such as breakdown strength, and improvement of thermal conductivity, tensile strength, and elasticity. Modulus and other thermal, mechanical properties, and nanofiller on the electric and thermal aging properties of the polypropylene composite material improvement effect is also very obvious.

At present, the research on the application of polypropylene and its nanocomposites in the insulation of high-voltage DC cables is still in its infancy, and there are still many issues that need further study:

1. A large number of experimental results show that nanodoping can improve the dielectric, thermal and mechanical properties of polypropylene, but scholars have not reached a consensus on the explanation of the improvement mechanism. A systematic and comprehensive analysis of the impact of factors such as the selection of single or multiple nanometers, the optimal ratio of additives, interface compatibility, nanosurface treatment, and dispersion

methods on the performance improvement of polypropylene is needed for the theoretical foundation research.

2. During normal operation, the insulation medium of high-voltage DC cables may also be affected by the coupling field consisting of stress strain field, electric field, magnetic field and temperature field. There is no corresponding experimental research on the dielectric, thermal and mechanical properties of polypropylene and its nanocomposites under coupled field.

3. Polypropylene materials are easily affected by electricity, heat, or light, aging, cable prepara- tion methods, process flow and other effects on the performance of polypropylene nanocom- posite insulation materials, the future still need to carry out corresponding research work.

Author details

Boxue Du, Zhaohao Hou and Jin Li*

*Address all correspondence to: lijin@tju.edu.cn

Key Laboratory of Smart Grid of Education Ministry, School of Electrical and Information Engineering, Tianjin University, Tianjin, China

References

[1] Fukawa M, Kawai T, Okano Y, et al. Development of 500-kV XLPE cables and accesso- ries for long distance underground transmission line. III. Electrical properties of 500-kV cables. IEEE Transactions on Power Delivery. 1996;**11**(2):627-634. DOI: 10.1109/61.329507

[2] He J, Chen G. Insulation materials for HVDC polymeric cables. IEEE Transactions on Dielectrics and Electrical Insulation. 2017;**24**(3):1307-1307. DOI: 10.1109/TDEI.2017.006721

[3] Paul DR, Robeson LM. Polymer nanotechnology: Nanocomposites. Polymer. 2008; **49**(15):3187-3204. DOI: 10.1016/j.polymer.2008.04.017

[4] Lewis TJ. Nanometric dielectrics. Insulation IEEE Transactions on Dielectrics & Electrical. 1994;**1**(5):812-825. DOI: 10.1109/94.326653

[5] Lewis TJ. Interfaces are the dominant feature of dielectrics at the nanometric level. IEEE Transactions on Dielectrics and Electrical Insulation. 2004;**11**(5):739-753. DOI: 10.1109/ TDEI.2004.1349779

[6] Tanaka T, Kozako M, Fuse N, et al. Proposal of a multi-core model for polymer nanocom- posite dielectrics. IEEE Transactions on Dielectrics and Electrical Insulation. 2005;**12**(4): 669-681. DOI: 10.1109/TDEI.2005.1511092

[7] Raetzke S, Kindersberger J. The effect of interphase structures in nanodielectrics. IEEE Transactions on Fundamentals & Materials. 2006;**126**(11):1044-1049. DOI: 10.1541/ieejfms. 126.1044

[8] Kurahashi K, Matsuda Y, Ueda A, et al. The application of novel polypropylene to the insulation of electric power cable. In: Transmission and Distribution Conference and Exhibition 2002: Asia Pacific. Yokohama, Japan: IEEE/PES; 2002. pp. 1278-1283

[9] Yoshino K, Ueda A, Demura T, et al. Property of syndiotactic polypropylene and its application to insulation of electric power cable-property, manufacturing and characteristics. In: Manufacturing and Characteristics Proceedings of the 7th International Conference on Properties and Applications of Dielectric Materials. Nagoya, Japan. 2003. pp. 175-178

[10] Hosier IL, Vaughan AS, Swingler SG. An investigation of the potential of polypropylene and its blends for use in recyclable high voltage cable insulation systems. Journal of Materials Science. 2011;46(11):4058-4070. DOI: 10.1007/s10853-011-5335-9

[11] Lee JH, Kim SJ, Kwon KH, Kim CH, Cho KC. A study on electrical properties of eco-friendly non-crosslinked polyethylene. In: 2012 IEEE International Conference on Condition Monitoring and Diagnosis, Bali. 2012. pp. 241-243. DOI: 10.1109/CMD.2012.64 16420

[12] Belli S, Perego G, Bareggi A, et al. P-laser: Breakthrough in power cable systems. In: Conference Record of the 2010 IEEE International Symposium on Electrical Insulation. San Diego, USA. 2010. pp. 1-5

[13] Fu M, Dissado LA, Chen G, et al. Space charge formation and its modified electric field under applied voltage reversal and temperature gradient in XLPE cable. IEEE Transactions on Dielectrics and Electrical Insulation. 2008;15(3):851-860. DOI: 10.1109/TDEI.2008.4543123

[14] Matsui K, Tanaka Y, Takada T, et al. Space charge behavior in low density polyethylene at pre-breakdown. IEEE Transactions on Dielectrics and Electrical Insulation. 2005;12(3): 406-415. DOI: 10.1109/TDEI.2005.1453444

[15] Suzuoki Y, Hattori K, Mizutani T, et al. Space charge and dielectric breakdown in polypropylene. In: IEEE, International Conference on Conduction and Breakdown in Solid Dielectrics. Leicester, UK: IEEE; 1995. pp. 641-645

[16] Montanari GC, Teyssedre G, Laurent C, et al. Investigating charge trapping behaviour of nanocomposite isotactic and syndiotactic polypropylene matrix. In: Proceedings of the 2004 IEEE International Conference on Solid Dielectrics. Vol. 2. Toulouse, France: IEEE; 2004. pp. 836-839

[17] Fuse N, Ohki Y, Tanaka T. Comparison of nano-structuration effects in polypropylene among four typical dielectric properties. IEEE Transactions on Dielectrics and Electrical Insulation. 2010;17(3):671-677. DOI: 10.1109/TDEI.2010.5492237

[18] Li Z, Cao W, Sheng G, et al. Experimental study on space charge and electrical strength of MgO nano-particles/polypropylene composite. IEEE Transactions on Dielectrics and Electrical Insulation. 2016;23(3):1812-1819. DOI: 10.1109/TDEI.2016.005181

[19] Abou-Dakka M, Chen Y. Effect of reverse polarity on space charge evolution in polypropylene with different concentration of natural and synthetic nano clay. In: IEEE

Conference on Electrical Insulation and Dielectric Phenomena. Shenzhen, China: IEEE; 2013. pp. 671-675

[20] Zhou Y, Hu J, Dang B, et al. Titanium oxide nanoparticle increases shallow traps to suppress space charge accumulation in polypropylene dielectrics. RSC Advances. 2016;**6**(54): 48720-48727. DOI: 10.1039/C6RA04868D

[21] Shimizu N, Laurent C. Electrical tree initiation. IEEE Transactions on Dielectrics and Electrical Insulation. 1998;**5**(5):651-659. DOI: 10.1109/94.729688

[22] Du BX, Zhu LW, Han T. Effect of low temperature on electrical treeing of polypropylene with repetitive pulse voltage. IEEE Transactions on Dielectrics and Electrical Insulation. 2016;**23**(4):1915-1923. DOI: 10.1109/TDEI.2016.7556462

[23] Holto J, Ildstad E. Electrical tree growth in extruded s-polypropylene. In: 2010 10th IEEE International Conference on Solid Dielectrics. Potsdam, Germany: IEEE; 2010. pp. 1-4

[24] Takala M, Karttunen M, Salovaara P, et al. Dielectric properties of nanostructured polypropylene-polyhedral oligomeric silsesquioxane compounds. IEEE Transactions on Dielectrics and Electrical Insulation. 2008;**15**(1):40-51. DOI: 10.1109/T-DEI.2008.4446735

[25] Takala M, Ranta H, Nevalainen P, et al. Dielectric properties and partial discharge endurance of polypropylene-silica nanocomposite. IEEE Transactions on Dielectrics and Electrical Insulation. 2010;**17**(4):1259-1267. DOI: 10.1109/TDEI.2010.5539698

[26] Virtanen S, Ranta H, Ahonen S, et al. Structure and dielectric breakdown strength of nano calcium carbonate/polypropylene composites. Journal of Applied Polymer Science. 2014;**131**(1):39504. DOI: 10.1002/app.39504

[27] Fabiani D, Montanari GC, Laurent C, et al. HVDC cable design and space charge accumulation. Part 3: Effect of temperature gradient [feature article]. IEEE Electrical Insulation Magazine. 2008;**24**(2):5-14. DOI: 10.1109/MEI.2008.4473049

[28] Couderc H, Fréchette M, David E. Fabrication and dielectric properties of polypropylene/silica nano-composites. In: 2015 IEEE Electrical Insulation Conference (EIC). Seattle, USA: IEEE; 2015. pp. 329-332

[29] Du BX, Cui B. Effects of thermal conductivity on dielectric breakdown of micro, nano sized BN filled polypropylene composites. IEEE Transactions on Dielectrics and Electrical Insulation. 2016;**23**(4):2116-2125. DOI: 10.1109/TDEI.2016.7556486

[30] Fukuyama Y, Kawai T, Kuroda SI, et al. The effect of the addition of polypropylene grafted SiO_2 nanoparticle on the crystallization behavior of isotactic polypropylene. Journal of Thermal Analysis and Calorimetry. 2013;**113**(3):1511-1519. DOI: 10.1007/s10973-012-2900-7

[31] Umemori M, Taniike T, Terano M. Influences of polypropylene grafted to SiO_2 nanoparticles on the crystallization behavior and mechanical properties of polypropylene/SiO_2 nanocomposites. Polymer Bulletin. 2012;**68**(4):1093-1108. DOI: 10.1007/s00289-011-0612-y

[32] Lee SH, Kontopoulou M, Park CB. Effect of nanosilica on the co-continuous morphology of polypropylene/polyolefin elastomer blends. Polymer. 2010;**51**(5):1147-1155. DOI: 10.1016/j.polymer.2010.01.018

[33] Du BX, Xu H, Li J, et al. Space charge behaviors of PP/POE/ZnO nanocomposites for HVDC cables. IEEE Transactions on Dielectrics and Electrical Insulation. 2016;**23**(5): 3165-3174. DOI: 10.1109/TDEI.2016.7736882

[34] Zhou Y, He J, Hu J, et al. Surface-modified MgO nanoparticle enhances the mechanical and direct-current electrical characteristics of polypropylene/polyolefin elastomer nanodielectrics. Journal of Applied Polymer Science. 2015;**133**(1):42863. DOI: 10.1002/app.42863

[35] Dang B, Zhou Y, He J, et al. Relationship between space charge behaviors and trap level distribution in polypropylene/propylene ethylene rubber/ZnO nanocomposites. In: 2016 IEEE Conference on Electrical Insulation and Dielectric Phenomena (CEIDP). Toronto, Canada: IEEE; 2016. pp. 595-598

[36] Guastavino F, Ratto A, Giovanna LD, et al. Syndiotactic polypropylene based nanocomposites: Short and long term electrical characterisation. In: 2012 Annual Report Conference on Electrical Insulation and Dielectric Phenomena (CEIDP). Montreal, Canada: IEEE; 2012. pp. 565-568

[37] Bulinski A, Bamji SS, Abou-Dakka M, et al. Dielectric properties of polypropylene loaded with synthetic organoclay. In: IEEE Conference on Electrical Insulation and Dielectric Phenomena. Virginia Beach, USA. 2009. pp. 666-671

The Performance of Insulation and Arc Interruption of the Environmentally Friendly Gas CF$_3$I

Dengming Xiao

Additional information is available at the end of the chapter

http://dx.doi.org/10.5772/intechopen.79968

Abstract

Many researches of trifluoroiodomethane (CF$_3$I) have shown that CF$_3$I has many excellent properties that make it one of the possible alternatives of SF$_6$. This paper reveals the effect laws of CF$_3$I gas content, gap distance, gas pressure, polarity, and electric field nonuniform coefficient on the insulation performance of CF$_3$I gas mixtures. In general, CF$_3$I-N$_2$ gas mixtures present a superior dielectric strength than CF$_3$I-CO$_2$ under different electric field sets. The experimental results indicate that 20 and 30% content CF$_3$I-N$_2$ gas mixtures can achieve nearly 50 and 55% insulation strength of pure SF6. In addition, to evaluate the arc interruption performance of environmentally friendly gas CF$_3$I, we set up a CF$_3$I transient nozzle arc model to study its thermodynamic and transport property. The analysis shows that CF$_3$I gas has a good arc interruption capability, which mainly functions thermodynamic and transport properties approach that of SF$_6$, and some are even better than SF$_6$. The decomposition process is also aggravated by impurities including metal and water. The main by-products are greenhouse gases with GWP below that of SF$_6$ and are lowly toxic and incombustible.

Keywords: gas insulation, environmentally friendly gases, CF$_3$I

1. Comparison of CF$_3$I and its mixtures with SF$_6$ and its mixtures on insulation property

In order to assess the insulation strength of CF$_3$I and its mixtures, power frequency breakdown voltages of SF$_6$ and 20%SF$_6$–80%N$_2$ mixture are measured in the same experimental condition [1]. The result is shown in **Tables 1** and **2**.

Gas	U_{SF_6} (kV)				$U_{20\%SF_6-80\%N_2}$ (kV)			
d (mm)/P (MPa)	0.3	0.2	0.15	0.1	0.3	0.2	0.15	0.1
5	126.8	82.0	60.7	34.9	82.2	53.7	39.0	25.3
10	222.3	150.2	112.9	82.9	170.5	112.6	82.5	57.7
15	315.0	220.1	174.8	120.0	246.0	166.7	123.9	90.7

Table 1. Power frequency breakdown voltages of SF_6 and 20%SF_6–80%N_2 gas mixtures in slightly nonuniform electric field.

Gas	U_{SF_6} (kV)				$U_{20\%SF_6-80\%N_2}$ (kV)			
d (mm)/P (MPa)	0.3	0.2	0.15	0.1	0.3	0.2	0.15	0.1
5	44.0	42.6	30.0	21.5	34.4	24.7	18.6	14.8
10	56.4	74.4	64.1	47.9	69.6	53.1	41.6	30.1
15	64.1	95.2	93.4	72.4	91.3	79.0	61.5	44.5
20	72.4	111.0	120.3	97.0	110.1	102.3	82.4	59.4

Table 2. Power frequency breakdown voltages of SF_6 and 20%SF_6–80%N_2 gas mixtures in highly nonuniform electric field.

Seen from **Table 1**, when P = 0.1 MPa and d = 10 mm, breakdown voltage of SF_6 in slightly nonuniform electric field is 82.9 kV, approximate to the standard value of breakdown voltage in uniform electric field, 89 kV/(mm•MPa). Moreover, breakdown voltage of 20%SF_6–80%N_2 mixture is 57.7 kV, roughly 70% of the value of SF_6, which is in agreement with data of other researchers.

Figure 1 compares the insulation property of CF_3I-N_2 gas mixtures [2] with SF_6 and SF_6-N_2 gas mixtures in slightly nonuniform electric field. It is obvious that the breakdown voltage of CF_3I-N_2 mixture is lower than SF_6 and SF_6-N_2 in the whole pressure range and distance range but higher than N_2 in the same condition. **Table 3** shows the insulation strength of CF_3I-N_2 mixture relative to pure SF6 and 20%SF_6–80%N_2 mixture in different distances and pressures. It can be known from **Table 3** that the insulation strength of 20%CF_3I–80%N_2 in slightly nonuniform electric field is as high as 50% of SF_6 and 70% of 20%CF_3I–80%N_2. If the mixing ratio of CF_3I increases to 30%, the insulation strength of CF_3I-N_2 mixture becomes 55% of SF6 and 78% of 20%SF_6–80%N_2. In this paper, CF_3I-CO_2 mixture, whose insulation strength is 97% of CF_3I-N_2 in the same condition if the mixing ratio is 30%, is researched [3]. With this proportion relationship, insulation property of CF_3I-CO_2 mixture relative to SF_6 and 20%SF_6–80%N_2 can be calculated [4].

About the situation in highly nonuniform electric field, breakdown voltage of CF_3I-N_2 mixture with the distances of 10 mm and 20 mm as examples is compared with SF_6 and 20%SF_6–80%N_2, as shown in **Figure 2**. Because of the strong electronegativity of SF_6 and CF_3I, the mixtures present drastic changing breakdown property. It is useless to define the relative insulation strength in highly nonuniform electric field generally, and analysis according to specific pressure condition is necessary [5].

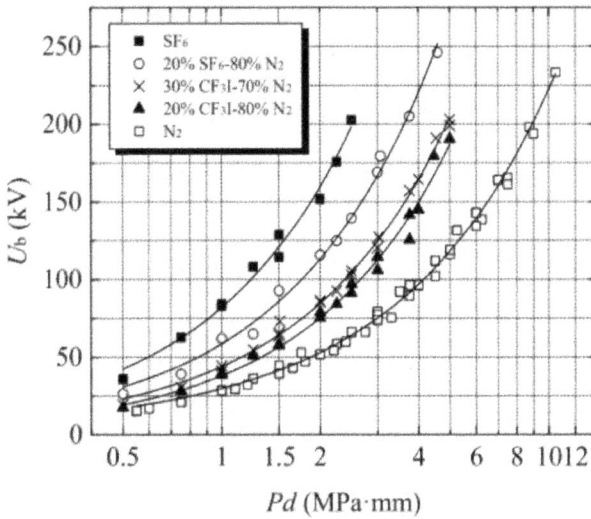

Figure 1. Comparison of the insulation property of CF_3I-N_2 mixture with SF_6 and SF_6-N_2 mixture in slightly nonuniform electric field.

P (MPa)	0.3	0.2	0.15	0.1	0.3	0.2	0.15	0.1
d (mm)	$U_{20\%CF_3I-80\%N_2}/U_{SF_6}$				$U_{20\%CF_3I-80\%N_2}/U_{20\%SF_6-80\%N_2}$			
5	0.53	0.49	0.49	0.50	0.82	0.75	0.77	0.67
10	0.54	0.52	0.51	0.48	0.70	0.70	0.70	0.69
15	0.57	0.52	0.48	0.49	0.73	0.69	0.68	0.64
d (mm)	$U_{30\%CF_3I-70\%N_2}/U_{SF_6}$				$U_{30\%CF_3I-70\%N_2}/U_{20\%SF_6-80\%N_2}$			
5	0.58	0.55	0.58	0.55	0.89	0.84	0.85	0.76
10	0.57	0.57	0.57	0.54	0.75	0.76	0.77	0.77
15	0.61	0.57	0.55	0.55	0.78	0.77	0.78	0.72

Table 3. Insulation strength of CF_3I mixture relative to SF_6 and 20%SF_6–80%N_2 in slightly nonuniform electric field.

Seen from **Figure 2**, power frequency breakdown voltage of SF_6 has obvious "hump effect" as a function of pressure with the distance of 10 and 20 mm. Similarly, the increase of breakdown voltage of 20%-mixed SF_6-N_2 mixture with pressure also tends to saturate. It is discovered by comparison that 30%CF_3I–70%N_2 possesses comparative insulation property to 20%SF_6–80%N_2, which is around 60% of SF_6. When $P = 0.2$ MPa, the insulation property of 30%CF_3I–70%N_2 mixture is as high as 80% of 20%SF_6–80%N_2 mixture and more than 55% of pure SF_6 [6]. When the pressure increases to 0.3 MPa, breakdown voltage of SF_6 declines rapidly because of the "hump effect." In this way, the breakdown voltage of 30%CF_3I–70%N_2 gas mixtures even exceeds pure SF_6, becoming 1.17 times higher than the latter. Because the power frequency breakdown voltage of 20%SF_6–80%N_2 always increases with the pressure, insulation strength

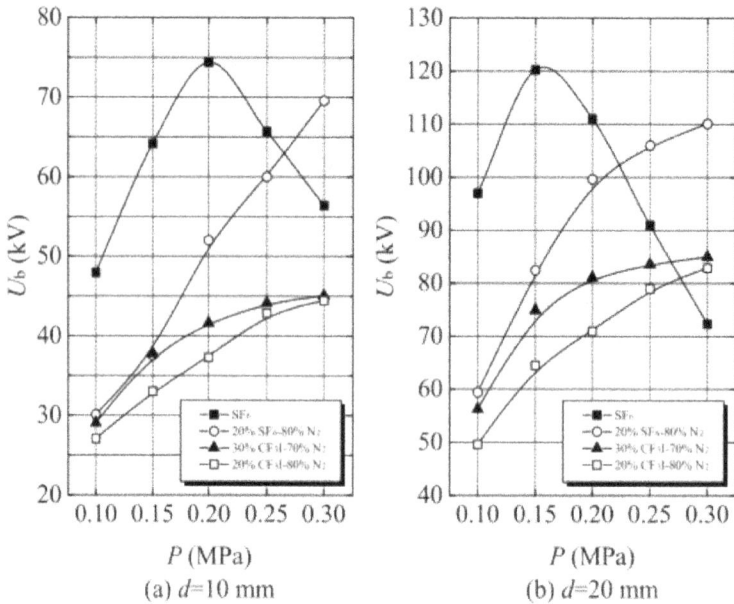

Figure 2. Comparison of the insulation property of CF_3I-N_2 gas mixtures with SF_6 and SF_6-N_2 mixture in highly nonuniform electric field.

of $30\%CF_3I$–$70\%N_2$ is only around 75% of $20\%SF_6$–$80\%N_2$ [7]. It can be indicated that the insulation strength of CF_3I-N_2 gas mixtures relative to SF_6-N_2 declines gradually with the increase of pressure, but the breakdown level is always more than 55% of SF_6. It is partly because the "hump effect" of SF_6 in the pressure is 0.1–0.3 MPa and also because interfusing buffer gas N_2 improves the abnormal breakdown phenomenon in highly nonuniform electric field [8]. **Table 4** shows the insulation strength of CF_3I gas mixtures relative to SF_6 and $20\%SF_6$–$80\%N_2$ in the same condition. It can be seen that the insulation property of $20\%CF_3I$–$80\%N_2$ in highly nonuniform electric field is more than 49% of SF_6 and 69% of $20\%SF_6$–$80\%N_2$ in the same condition.

CF_3I-CO_2 gas mixtures, also affected by "hump effect," present obvious decline in high pressure. Under the pressure 0.1 and 0.15 MPa, the breakdown voltage of 30%-mixed CF_3I-CO_2 gas mixtures is 60% of pure SF_6 and 1.05 times of $20\%SF_6$–$80\%N_2$ gas mixtures [9]. However when the pressure rises to 0.3 MPa, the insulation property of CF_3I-CO_2 gas mixtures is only around 40% of $20\%SF_6$–$80\%N_2$ mixture [10]. Relative insulation strength of different pressures is shown in **Table 4** in detail.

The insulation properties of CF_3I-N_2 and CF_3I-CO_2 gas mixtures can be concluded with the analysis above.

In uniform electric field, the insulation strength of 30%-mixed CF_3I-N_2 gas mixtures can approach 72% of pure SF_6 theoretically and in slightly nonuniform electric field more than 55% of pure SF_6 [11]. In highly nonuniform electric field, the relative insulation strength of

d (mm)	10	20	10	20
P (MPa)	$U_{30\%CF_3I-70\%N_2}/U_{SF_6}$		$U_{30\%CF_3I-70\%N_2}/U_{20\%SF_6-80\%N_2}$	
0.10	0.61	0.58	0.97	0.95
0.15	0.59	0.62	0.91	0.91
0.20	0.55	0.73	0.77	0.79
0.30	0.80	1.17	0.75	0.77
P (MPa)	$U_{20\%CF_3I-80\%N_2}/U_{SF_6}$		$U_{20\%CF_3I-80\%N_2}/U_{20\%SF_6-80\%N_2}$	
0.10	0.56	0.51	0.90	0.84
0.15	0.51	0.54	0.79	0.78
0.20	0.49	0.64	0.68	0.69
0.30	0.79	1.15	0.69	0.75
P (MPa)	$U_{30\%CF_3I-70\%CO_2}/U_{SF_6}$		$U_{30\%CF_3I-70\%CO_2}/U_{20\%SF_6-80\%N_2}$	
0.10	0.64	0.62	1.02	1.02
0.15	0.62	0.59	1.05	0.86
0.20	0.55	0.42	0.77	0.46
0.30	0.67	0.59	0.55	0.39

Table 4. Insulation strength of CF$_3$I gas mixtures relative to SF$_6$ and 20%SF$_6$–80%N$_2$ gas mixtures in highly nonuniform electric field.

CF$_3$I-N$_2$ gas mixtures is influenced by the pressure seriously, which ranges from 55 to 117% in the researched pressure range [12].

The insulation strength of 30%-mixed CF$_3$I-CO$_2$ gas mixtures can approach 68% of pure SF$_6$ theoretically in uniform electric field, and in slightly nonuniform electric field, it can be more than 53% [13]. In highly nonuniform electric field, the relative insulation strength is also influenced by the pressure. The insulation strength of 30%-mixed CF$_3$I-CO$_2$ gas mixtures is 0.42–0.62 times of pure SF$_6$.

Detailed comparative data and the insulation strength of 20%SF$_6$–80%N$_2$ gas mixtures are listed in **Table 5**. Seen from the table, whether in slightly nonuniform electric field or highly nonuniform electric field, the insulation strength of CF$_3$I-N$_2$ is better than CF$_3$I-CO$_2$ in the same condition. Therefore, in actual equipment application, CF$_3$I-N$_2$ gas mixtures with the mixing ratio of 20–30% should be considered preferentially.

Then, principles of applying CF$_3$I mixture in actual electrical equipment are discussed with the example of 40.5 kV cubicle-type gas-insulated switchgear (C-GIS).

C-GIS, vulgarly named "gas-filled cabinet," is suitable for situations that have high requirement on reliability and limited space and floor area such as rail transit, high buildings, and industrial enterprises. 40.5 kV C-GIS is the most popularly employed. Known from the survey

Gas	Electric field	$U_{CF_3I\text{-}x}/U_{SF_6}$	$U_{CF_3I\text{-}x}/U_{SF_6\text{-}N_2}$
	Uniform	0.72	0.88
30%CF$_3$I–70%N$_2$	Slightly nonuniform	0.55	0.78
	Highly nonuniform	0.55–1.17	0.75–0.97
	Uniform	0.68	0.83
30%CF$_3$I–70%CO$_2$	Slightly nonuniform	0.53	0.75
	Highly nonuniform	0.42–0.67	0.39–1.02

Table 5. Insulation strength of CF$_3$I gas mixtures relative to SF$_6$ and 20%SF$_6$–80%N$_2$ gas mixtures.

of products on the market, the most popular practice currently is employing SF$_6$ with the pressure 0.12–0.15 MPa (abs) as the main insulation medium of primary loop. For example, the SF$_6$ charge pressures of XGN46–40.5 model C-GIS developed by Xi'an High Voltage Apparatus Research Institute and ZX2 model C-GIS produced by ABB are, respectively, 0.15 and 0.13 MPa. Meanwhile, some companies employ higher pressure, e.g., 8DA10 model C-GIS produced by Siemens, whose SF$_6$ charge pressure is 0.26 MPa. C-GIS is complicated on structure and different on shape, containing function units such as breakers, disconnectors, current/voltage transformers, etc., and forming both slightly nonuniform electric field and highly nonuniform electric field. In accordance to relative insulation strength shown in **Table 5**, replacing SF$_6$ in C-GIS with CF$_3$I-N$_2$ with the mixing ratio of 20–30% as insulation medium without changing the gap distance and pressure, insulation strength is not strong enough, so it is necessary to adjust the charge pressure or structure properly.

Through analysis, we can know that increasing pressure or increasing gap distance can effectively improve the insulation level of CF$_3$I gas mixtures. If the gap distance is kept constant, the CF$_3$I-N$_2$ gas mixtures of 20–30% mixing ratio at 0.25–0.3 MPa will reach the insulation level of SF$_6$ at 0.15 MPa. In this case, there is no need to change the existing C-GIS internal structure; only a thickened metal enclosure is needed to ensure the operation pressure and the leakage level meet the requirements. At the same time, the liquefaction temperature of CF$_3$I-N$_2$ gas mixtures around this pressure can be maintained below −25°C, which is in accordance with the requirements of GB/T 11022-2011 on the operating temperature of the switch equipment. If the operating pressure is kept constant, the data in **Figure 1** and **Table 1** show that to achieve the same insulation strength as SF$_6$, the gap distance of 30%CF$_3$I–70%N$_2$ gas mixtures needs to be increased by about twice as much.

Therefore, although the insulation strength of the CF$_3$I gas mixtures is weaker than pure SF$_6$, CF$_3$I gas mixtures can not only achieve the same insulation level as the SF$_6$, but also the liquefaction temperature can meet the operation requirement of the switch equipment by adjusting the operating condition or structure parameters properly. More importantly, CF$_3$I-N$_2$ gas mixtures will not affect global warming. It can solve the unfriendly environmental problems of SF$_6$ and realize the green upgrading of electrical equipment. Therefore, it is suggested that CF$_3$I-N$_2$ gas mixtures with the mixing ratio of 20–30% can be used as alternative medium for SF$_6$ in low- and medium-voltage electrical equipment.

Based on the results of power frequency and lightning impulse tests, it is shown that the insulation strength of CF_3I-N_2 gas mixtures increases with the increase of CF_3I mixing ratio in slightly nonuniform electric field or in highly nonuniform electric field. For CF_3I-CO_2 gas mixtures, the breakdown voltage varies with the mixing ratio in slightly nonuniform electric field, similar to CF_3I-N_2. However, in highly nonuniform electric field, the breakdown voltage of 10%-mixed CF_3I-CO_2 gas mixtures is higher than mixtures with mixing ratio of 20 and 30%, due to the abnormal breakdown of the high pressure. It can be seen that higher mixing ratio of electronegative gas in the CF_3I gas mixtures does not always cause better insulation performance, and it will also be influenced by many factors such as the gap distance, the pressure, and the unevenness of the electric field.

From the change of the breakdown voltage with the gap distance, the breakdown voltage of CF_3I-N_2 and CF_3I-CO_2 gas mixtures increases linearly with the increase of the gap, and the increase speed is proportional to the pressure in slightly nonuniform electric field. In highly nonuniform electric field, the breakdown voltage of CF_3I-N_2 gas mixtures shows a tendency to saturate with the gap distance, while the CF_3I-CO_2 combination has obvious nonlinear characteristics.

From the change of the breakdown voltage with the pressure, the insulation strength of the CF_3I-N_2 and CF_3I-CO_2 gas mixtures in the slightly nonuniform field increases linearly with the increase of the pressure and even shows a certain degree of negative synergy. However, under the highly nonuniform electric field environment, there is a clear "hump" in the power frequency breakdown voltage of CF_3I-CO_2 gas mixtures with the change of pressure, and the positive lightning impulse coefficient in the "hump" section is always less than 1.

From the effect of polarity on discharge, the breakdown voltage of CF_3I gas mixtures under positive polarity is higher than that of negative polarity in slightly nonuniform electric field. But in highly nonuniform electric field, the opposite is true. In addition, the uniformity of electric field will also affect the insulation performance of CF_3I gas mixtures. Therefore, in the actual product application, the uniformity of the electric field should be improved as much as possible for the purpose to ensure that the insulation strength of the CF_3I gas mixtures can be maintained at a high level.

Finally, by comparing with the SF_6 and 20%SF_6–80%N_2, it is found that the 20%CF_3I–80%N_2 gas mixtures can reach the insulation level of 50% of pure SF_6 gas and about 65% of 20%SF_6–80%N_2 gas mixtures. When the mixing ratio of CF_3I is increased to 30%, insulation strength of CF_3I-N_2 gas mixtures can reach about 55% and 75% of SF_6 and 20%SF_6–80%N_2 gas mixtures, respectively. For the combination of CF_3I-CO_2, 30%-mixed CF_3I-CO_2 gas mixtures can reach the insulation level of more than 53% of pure SF_6 in slightly nonuniform electric field, but in highly nonuniform electric field, relative insulation strength depends on the pressure and can only reach 42–67% of the latter.

2. The radial temperature distribution characteristics and leading energy transport process of CF_3I nozzle arc

The transient temperature distribution of arc is decided by the mutual equilibrium among the energy input of arc, different energy transport processes, and changing rate of energy storage, which determines the characteristics of arc in turn. We analyze the arc extinction characteristics of CF_3I through the thermodynamic characteristics of CF_3I arc.

The temperature of arc plasma changes with the changes of time and position, and that's decided by the interactions of the produce of the joule heat of arc and different energy transport processes in itself. The following are the calculation and analysis of the energy balance of the core area (in the radius with 83.3% highest temperature) and conducting regions (in the radius with 4000K temperature) of arc in the conditions with 900 A large current, 50 A small current, and current-zero period.

By the radial temperature analysis of arc, during the process that the head of the moving contact moves from the nozzle upstream to the throat, because the space of the throat is small, and the moving contact and thermal plasma of arc will block the flow of the fluid to the catchment area through the nozzle, and the slow speed of the fluid causes a bigger radius of arc so that the core area is not obvious and the temperature is relatively lower. We choose the time of 900 A current to analyze in **Table 6**. In the time of large current, the sinusoidal current waveform changes relatively slowly with the change of time, and the changing rate of energy storage of the core area of arc can almost be negligible, and as the slow decline of arc, global energy of arc also goes down, so the change of energy storage in the radius of arc reaches more than 20%. As the interior of the thicker arc cannot be influenced by the cold fluid, the electric power produced in the core area of arc is lost more than 90% by the radiation process, and beyond the core area of arc, the interactions of arc and fluid are more drastic, and the energy loss of radiation lowers obviously. As a whole, the axial convection process of arc does positive work, and the core area does less work. Although the radial convection of arc is small, it takes the joule heat away, and that's because the cold fluid in upstream moves from the entrance to the lower right and contacts with the thermal plasma with a bigger oblique angle to form convection, especially the cold fluid which flows along the slope of fixed contact even forms eddy in the interior of arc. Therefore, the radial convection in the arc boundary has a good cooling effect, especially in the reabsorption area of the radiation energy.

When the current lowers to 50 A, the moving contact moves to the catchment of nozzle, and the distance of contact reaches more than 18 mm. The throat of the nozzle opens completely, and the thermal characteristics of the nozzle arc are decided by the energy distribution and transport process in **Table 7**. Differing from large current, because of the temperature distribution

Arc boundary	Electric power input (10^3W)	Radiation loss (%)	Radial thermal conductance (%)	Axial thermal convection (%)	Radial thermal convection (%)	Energy loss of pressure to do work (%)	Changing rate of energy storage of electric power (%)
Boundary of core area (R_{833})	4.185	−93.9	−14.0	13.6	−11.2	−0.1	2.4
Arc boundary (R_{4k})	4.768	−25.8	−52.6	25.8	−68.7	0.2	21.9

Table 6. Percentage of electrical power input associated with various energy transport processes for the whole CF_3I arc length at core and arc boundary at 900 A.

Arc boundary	Electric power input (10^3W)	Radiation loss (%)	Radial thermal conductance (%)	Axial thermal convection (%)	Radial thermal convection (%)	Energy loss of pressure to do work (%)	Changing rate of energy storage of electric power (%)
Boundary of core area (R_{833})	4.179	−46.6	−69.4	17.8	−6.2	−10.5	11.7
Arc boundary (R_{4k})	5.666	−0.1	−79.4	−25.7	21.0	−16.2	4.7

Table 7. Percentage of electrical power input associated with various energy transport processes for the whole CF₃I arc length at core and arc boundary at 50 A.

characteristics of arc and the effect of high-speed cold fluid, the loss ratio of radiation energy decreases sharply. After the reabsorption process of radiation in the arc boundary, the loss ratio of radiation energy approaches zero. Around arc, the high-speed movement of the cold fluid in the nozzle makes arc form an obvious high-temperature core area. The fast change of radial temperature makes the radial thermal conduction become strong, and the energy transport ratio in the core area and arc area is both 70%. In comparison, the convection effect works little to the change of energy. Because the moving contact pulls to the right and arc plasma strengths in axial, the axial convection does positive works in the core area of arc, but in the arc boundary, the high-speed cold fluid around arc effectively takes the energy away, and it has an obvious cooling effect to arc. The radial convection in the core area begins to do negative works, and that's because the temperature gradient in the core area of arc is very big, and the gas with relatively lower temperature in the radiation reabsorption region begins to enter into the core area to maintain the conservation of mass in the interior of arc; however, out of the core area of arc, the decline of the radial temperature becomes slow, and the radial convection effect still does positive works.

In the current-zero area, shown in **Table 8**, the joule heat of arc is zero, and the temperature of plasma lowers to less than 10,000 K, and the radiation effect is almost 0. The radius of arc

Arc boundary	Electric power input (10^3W)	Radiation loss (%)	Radial thermal conductance (%)	Axial thermal convection (%)	Radial thermal convection (%)	Energy loss of pressure to do work (%)	Changing rate of energy storage of electric power (10^3W)
Boundary of core area (R_{833})	0	−0.1	−13.3	−35.8	−50.4	−2.4	1.796
Arc boundary (R_{4k})	0	−0.0	−19.0	−37.4	−48.7	−1.8	1.841

Note: Because the current input in the current-zero area is zero, the change of arc's energy storage is the changing power of energy, and it is regarded as the energy input to measure the strength of every energy transport process.

Table 8. Percentage of electrical power input associated with various energy transport processes for the whole CF₃I arc length at core and arc boundary at current zero.

is very small, and the thermal conduction is relatively weaker. Convection mainly consumes arc's energy, whether in the core area or within the arc boundary, and the radial thermal convection takes more than 30% energy away as the axial reaches about one half of the change of arc's energy.

3. The arc's characteristics near the arc-zero area

The thermal interruption of nozzle arc is decided by the arc's temperature around the current-zero area. Through the discussion of the above section, when it's close to the current-zero area (25 A and lower), the arc's temperature and radius decrease rapidly and become a key time of arc interruption, and this process is worth being researched more. When the current is very small, a strong radial thermal conduction makes arc form a relatively bigger radial temperature gradient in the radiation reabsorption area, which takes the energy in the arc's interior away effectively. And this promotes more entrance of cold fluid of outside into the high-temperature plasma of arc (to maintain the conversation of mass), so that the radial convection heat dissipation becomes the main form of energy transport when the current approaches zero. The superposition of two effects explains the sharp declines of the arc's temperature and radius in dozens of microseconds before current zero-crossing in **Figures 3** and **4**.

The changes of nozzle throat arc's temperature and radius with time near current-zero area are shown in **Figures 3** and **4**. They show that in 30 s before arc-zero area, CF_3I arc's temperature shows a trend of faster decrease without obvious phenomenon that the declines of temperature about time decrease. And the arc's radius decreases with time. From the characteristic

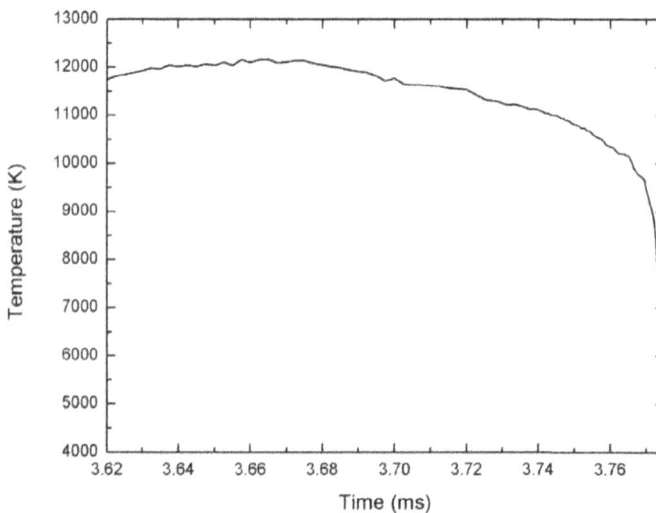

Figure 3. Arc temperature varied with time at nozzle throat near current zero.

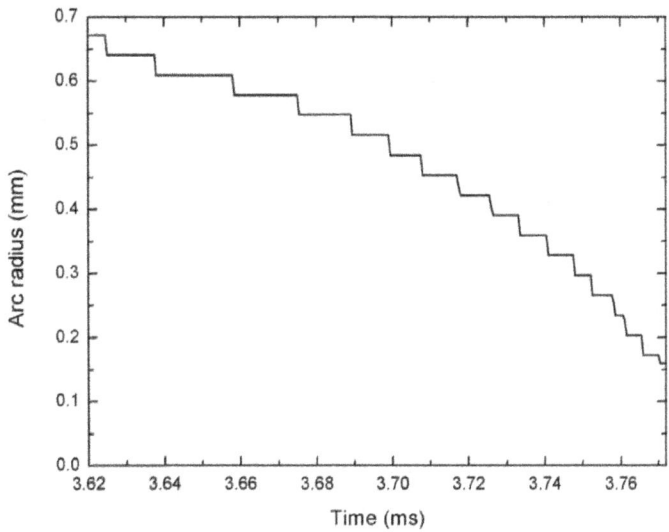

Figure 4. Arc radius varied with time at nozzle throat near current zero.

	About arc's radius	About arc's temperature
10 μs before current zero-crossing	24.75 μs	40.64 μs
Current-zero area	3.90 μs	2.27 μs

Table 9. Characteristic time of CF$_3$I arc near current zero.

time of arc (see **Table 9**), at 10 μs before current zero-crossing (arc current is about 5 A), the characteristic time about arc's radius and temperature is 24.75 and 40.64 μs which are equivalent to SF$_6$ arc's values in the same time. In the last 10 μs, the characteristic time of CF$_3$I arc in the current-zero area lowers to about 3 μs which is much smaller than air and CO$_2$ arc and approaches SF$_6$ arc. The comparison shows that the declines of CF$_3$I arc's temperature and radius near current-zero area are equivalent to SF$_6$ and much bigger than air and CO$_2$ arc.

4. Analysis of the by-product after CF$_3$I interrupts arc

There are three reasons to cause gas decomposition, and they are decomposition caused by electron collision, thermal decomposition, and photodecomposition. There are mainly first two types in the high-voltage electrical equipment. The main discharge forms in switching arc or equipment such as GIS are high-power arc discharge, spark discharge, and partial discharge. Among them, high-power discharge has large current and a long time of duration, and energy can accumulate in a short time, and temperature can reach more than 20,000 K, and it's the most serious discharge form of decomposition reaction.

Although physical parameters of CF_3I show good thermal interruption characteristics, the same with SF_6, it will have decomposition reaction in the conditions of high temperature and high gas pressure. Though most decomposition reactions are reversible reaction, after arc interrupt, there are still some by-products to be produced. These by-products will influence subsequent quenching of arc, and their toxicity and environmental characteristics will influence the application of CF_3I in switching arc equipment. This section does spectrum analysis to the arc-later gases, combining the calculating results of the thermomotive equilibrium state gas composition, and evaluates the types of arc-later by-products of CF_3I and the possible influence from the impurities.

4.1. Spectrum analysis to the arc-later compositions of CF_3I

The arcing experiment is done in the SF_6 switch cabinet with nominal voltage of 12 kV and nominal current of 630 A, and we use the sine-input arc current of effective value of 400 and 630 A. We conduct sampling of the arc-later gas of CF_3I after five times' arcing experiment. The experimental gas is CF_3I with purity of more than 99.5% which has known impurities of CO_2 and CF_3Br and water with volume ratio lower than 18 ppm (1.8×10^{-3}%). Before the experiment, we have done the insulation test and moisture content test to the switch cabinet to ensure that it can reach experimental standard.

Gas detection uses PLOT-Q chromatogram column to separate and detect the gas, and the detection is done 48 hours after sampling of gas. We use the cleansed needle tubing for sampling and put into the detection equipment for automatic detection, gas separation, and spectrum distinction. Spectrometer has a distinguishing uniformity of more than 80% to the gas composition, and we can trust the judgment of gas composition. As needle tubing sampling will inevitably mix with air, being a mass of N_2 and trace amounts of CO_2, we will find N_2 and CO_2 in the detection. Every gas sample is about 10 μL.

Arc-later gas composition gotten by spectral testing is shown in **Table 10**. In the experiment, 400 A current successfully interrupts for five times, but 630 A current restrikes in the first interruption experiment and continues to burn for dozens of milliseconds. Therefore, arcing

	Original gas	400 A five times	630 A one time
N_2	√	√	√
CF_4	√	√	√
CO_2	√	√	√
C_2F_6		√	
CHF_3	√	√	
H_2O	√	√	√
C_4F_8	√	√	
CF_3I	√	√	√

Table 10. Gas composition after arc for CF_3I arc obtained through mass spectrum measurement.

experiment under 630 A current is just done for one time, and we obtain the gas after unsuccessful interruption as testing sample. In the results, except the main composition CF_3I, other impurities or gas decompositions are N_2, CF_4, CO_2, C_2F_6, CHF_3, H_2O, and C_4F_8. N_2 and H_2O are impurities because of mixture of air. CO_2, CHF_3, and C_4F_8 exist in the original gas, so they are the impurities in the producing process of gas. Therefore, CF_4 and C_2F_6 are the products of the high-voltage and large-current arcing experiment of CF_3I. The comparisons of the response values of CF_3I, CF_4, C_2F_6, C_4F_8, and CHF_3 in original gas and after different times' arcing experiment are shown in **Figure 5**.

From the spectrum data of CF_3I, CF_3I decomposes little after five times' successful interruption of 400 A arc, and CF3I is 97.0% of original gas; the decomposition can be ignored. But after the unsuccessful interruption of 630 A current, CF_3I is only 3.4% of original gas. Other gases' spectrum data shows that under conditions of successful interruption, only 3% CF_3I produces CF_4 and C_2F_6 and little C_4F_8 and CHF_3 as the gas reaches room temperature from arc's high temperature. According to conservation of elements, it will separate out a little iodine. When the arcing interruption fails, more than 95% CF_3I becomes CF_4 and separates out iodine at the same time. This shows that the ratio of by-product after successful arcing interruption of CF_3I is very low and the main by-products are CF_4 and C_2F_6. However, when arc interruption fails and arc continues to burn, a mass of CF_3I will decompose and produce much CF_4 so that it cannot be used as arc-quenching medium. By examining the switch cabinet after the failure of arc quenching, we can find that there are much black solid attached on the surface of insulator and contact and this solid is carbon and iodine. The production of elementary substance shows that during the drastic burning of arc-restrike gas, many organic compounds decompose into elementary substance due to the lack of oxidizer (such as oxygen). Therefore, when CF_3I is applied in interrupting arc, we should reserve enough margin for interrupting current, and at the same time, we can use some absorbents or make air-blast-arc method to eliminate produced iodine.

4.2. Characteristic analysis of CF₃I decomposing products

Based on environmentally friendly purpose and toxicity, we analyze the composition of arc-later gases, and the results are listed in **Table 11**. It shows that compared with SF_6, the global warming potential (GWP) of decomposing gases of CF_3I under conditions of large current decreases at different levels. Except CF_4, the existence time in the atmosphere of other gases drastically reduces. As for toxicity, the arc-later decompositions of CF_3I belong to perfluorocarbon and partial fluorocarbon, which both have low toxicity, so it will cause headache, nausea, or dizziness under conditions of long-time or high-density inhalation. While doing experiment, we should note ventilation and appropriate self-protection.

In a word, after CF_3I arc discharges under conditions of high voltage and large current, main by-products are perfluorocarbons, CF_4 and C_2F_6; in the event of mixing trace amounts of water, it will produce a little hydrofluorocarbons such as CHF_3 and C_2HF_5. Perfluorocarbon and hydrofluorocarbon with few carbon atoms both have low toxicity; the consequences of inhalation are connected with the density. Only in closed environment, when human inhales higher-density gas, it will damage the cardiovascular system to a certain degree. During experiment and use, we should note ventilation and appropriate self-protection to ensure a safe process.

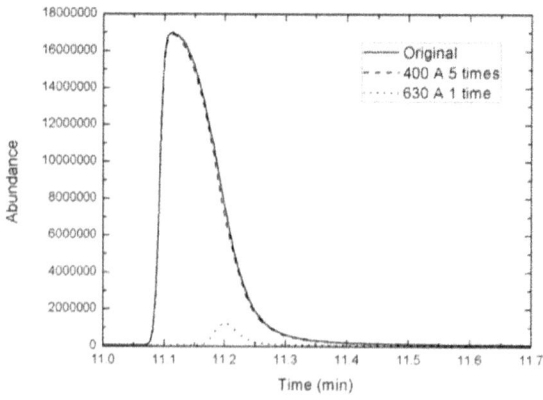

(a)

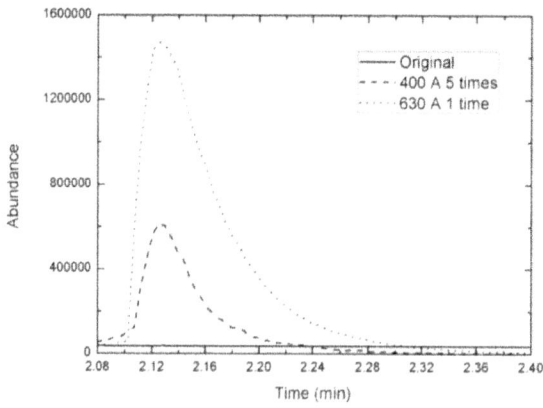

(b)

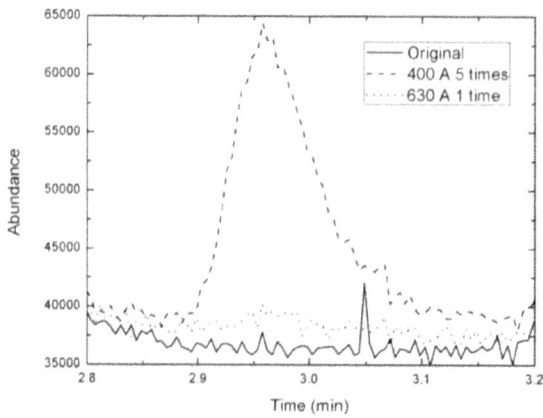

(c)

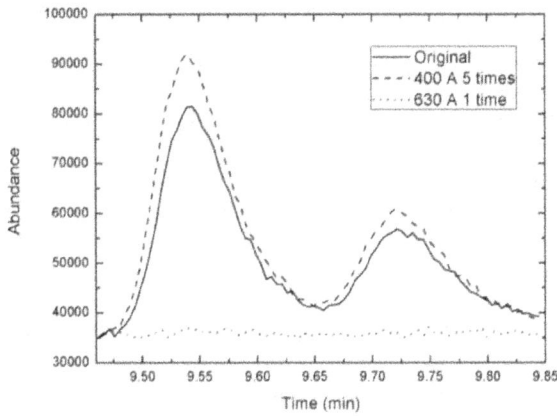

(d)

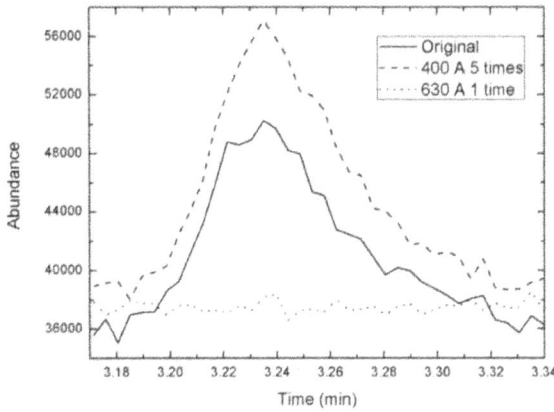

(e)

Figure 5. Mass spectrum data of CF₃I gas composition after arcs: (a) CF₃I, (b) CF₄, (c) C₂F₆, (d) C₄F₈, and (e) CHF₃.

Gas	Dielectric strength compared with SF₆	Liquefaction temperature (°C)	GWP	Existence time in the atmosphere (year)	Toxicity
C₂F₆	0.78–0.79	−78	9200	10,000	Low toxicity
CF₄	0.39	−186.8	6500	50,000	Low toxicity
CHF₃	0.18	−78.2	11,700	264	Low toxicity
C₂HF₅	0.59	−48.5	14,800	32.6	Low toxicity

Table 11. Environmental and toxic analysis of CF₃I gas composition after arc.

5. Conclusions

1. After adjustment to operation conditions or structure size of the electrical equipment, CF_3I gas mixtures can reach equivalent insulation level of SF_6, and at the same time, the liquefaction temperature can satisfy operation conditions of switch equipment. Because CF_3I-N_2 gas mixtures will not influence global warming, it can resolve environmentally unfriendly problems and realize the green upgrade of the electrical equipment. Therefore, we suggest that 20–30% CF_3I-N_2 gas mixtures can be applied as SF_6 alternatives in medium- and low-voltage electrical apparatus.

2. CF_3I has a good arc interruption characteristic, and before current zero-crossing, it approaches SF_6 and some thermodynamic properties are even better than SF_6. CF_3I decomposes easily after large-current arcs, high-temperature decomposing products are hard to recombine after arc extinction, and they are easy to be influenced by impurities such as water and so on and produce a little hydrofluorocarbon. Few-carbon-atom perfluorocarbon and hydrofluorocarbon both have low toxicity, so we should take appropriate actions such as absorbent or gas mixtures to eliminate or restrain decomposing products.

Acknowledgements

This work is supported by the National Natural Science Foundation of China (Grant No.51337006).

Author details

Dengming Xiao

Address all correspondence to: dmxiao@sjtu.edu.cn

Department of Electrical Engineering, Shanghai Jiao Tong University, Shanghai, China

References

[1] Taki M, Maekawa D, Odaka H, Mizoguchi H, Yanabu S. Interruption capability of CF_3I gas as a substitution candidate for SF_6 gas. IEEE Transactions on Dielectrics and Electrical Insulation. 2007;14:341-346

[2] Cressault Y, Connord V, Hingana H, Teulet P, Gleizes A. Transport properties of CF_3I thermal plasmas mixed with CO_2, air or N_2 as an alternative to SF_6 plasmas in high-voltage circuit breakers. Journal of Physics D: Applied Physics. 2011;44:495202

[3]   Wang W, Rong M, Wu Y, Yan JD. Fundamental properties of high-temperature SF$_6$ mixed with CO$_2$ as a replacement for SF$_6$ in high-voltage circuit breakers. Journal of Physics D: Applied Physics. 2014;**47**:255201

[4]   Liu J, Zhang Q, Yan J, Zhong J, Fang M. Analysis of the characteristics of DC nozzle arcs in air and guidance for the search of SF$_6$ replacement gas. Journal of Physics D: Applied Physics. 2016;**49**:435201

[5]   Zhang X, Dai Q, Han Y, et al. Investigation towards the influence of trace water on CF$_3$I decomposition components under discharge. High Voltage Technology. 2016;**42**:172-178

[6]   Yokomizu Y, Ochiai R, Matsumura T. Electrical and thermal conductivities of high-temperature CO$_2$–CF$_3$I mixture and transient conductance of residual arc during its extinction process. Journal of Physics D: Applied Physics. 2009;**42**:215204

[7]   Yokomizu Y, Ochiai R, Matsumura T. Particle composition of CO$_2$-CF$_3$I mixture at temperatures of 300-30,000 K. The Transactions of the Institute of Electrical Engineers of Japan B. 2007;**127**:1281-1286

[8]   Qiu Y. GIS Equipment and Its Insulation Technologies. Beijing: China Water & Power Press; 1994

[9]   Xiao S, Cressault Y, Zhang X, Teulet P. The influence of Cu, Al, or Fe on the insulating capacity of CF$_3$I. In: Physics of Plasmas (1994-present), vol. 23. 2016. p. 123505

[10] Christophorou LG, Olthoff JK, Green DS. Gases for Electrical Insulation and Arc Interruption: Possible Present and Future Alternatives to Pure SF6. NIST TN-1425, vol. 82011. p. 391

[11] Deng Y. Fundamental research of the application of environmentally friedly insulation gas CF$_3$I in the electrical equipment [PhD thesis]. Shanghai: Shanghai Jiao Tong University; 2016

[12] Christophorou LG. Insulating gases. Nuclear Inst & Methods in Physics Research A. 1988;**268**:424-433

[13] Zhao X. The research of arc extinguishing and insulation characteristics of environmentally friendly insulation gas CF$_3$I [PhD thesis]. Shanghai: Shanghai Jiao Tong University; 2018

Analysis for Higher Voltage at Downstream Node, Negative Line Loss and Active and Reactive Components of Capacitor Current, and Impact of Harmonic Resonance

Sushanta Paul

Additional information is available at the end of the chapter

http://dx.doi.org/10.5772/intechopen.80879

Abstract

In a study, it was found that the voltage at the downstream node is higher than the voltage at the upstream node, even though all the current flows from the upstream node to the downstream node. In IEEE's load flow simulation results for the 13-bus system, 34-bus system, and 123-bus system, it was also found that line losses in some feeders are negative. In this chapter, it has been analyzed how higher voltage at the downstream node and negative line losses in a phase appear in an AC power system. It has also been demonstrated that even though a capacitor generates only reactive power, its current has both active and reactive components with respect to the system reference. Finally the impact of harmonic resonance on capacitor has been discussed.

Keywords: capacitors, line loss, load flow, power system simulation, reactive power

1. Introduction

We generally know that current flows from high voltage to low voltage and this is true for a DC system but that does not apply to an AC system, because it can be seen that even though current is flowing from upstream node to downstream node, voltage at downstream node is higher than the voltage at upstream node. The reason of having higher voltage at downstream node in an AC system is that voltage drop in a single phase depends on mutual impedance and current in other phases besides self-impedance and current of its own phase. In [1], Vienna rectifier (three-phase/three-switch converter) and Z-source inverter (ZSI) have been proposed with fewer number switches to boost the DC voltage and reduce the voltage sag.

Load flow simulation results for 13-bus system, 34-bus system, and 123-bus system show that line losses in some feeders are negative. Negative line losses may appear in the lighted loaded phase in unbalanced system. Distribution system is practically unbalanced system because of unbalanced loads and having single-phase and two-phase lines. From the operational stand-point, unbalanced conditions typically occur in the normal operation of aggregated loads or during short periods of abnormal operation with unbalanced faults or with one/two phases out of service [2].

Generally there are two approaches to measure the line losses. One is classical approach where line losses are computed as the difference between input power at upstream node and output power at downstream node; in this approach, losses in neutral and dirt are included in each single phase of a three-phase line. In other approaches, line losses are computed as (I^2R), phase resistance multiplied by current squared. In this second approach, losses in neutral and dirt need to be calculated separately to determine the total three-phase losses. In the study, for example, the IEEE 13-bus system, it is seen that some single-phase lines have negative line losses and that negative line losses show up in the classical approach. It needs to be mentioned that, although single-phase line losses may appear as negative in the first approach, total three-phase losses are the same as total three-phase losses determined in the second approach and these have been shown in [3]. Even though negative line losses appear in classical approach-based line loss computation, there is no physical explanation of negative line losses where positive line losses are considered as electrical energy that dissipates as heat energy when electric current flows through the line. In [2], Carpaneto et al. proposed a resistive component-based loss partitioning (RCLP) method. Their results show that even though single-phase line losses obtained by RCLP method differ from line losses obtained by classical approach, three-phase line losses obtained by RCLP method are the same as three-phase line losses obtained by classical approach. In our previous work [4], it has been shown that line losses increase or decrease at reduced voltage, depending on the types of the loads. In [5], it was demonstrated how temperature, depending on the type of the load, influences the variations in line losses.

A capacitor has huge application in the power system for reactive power compensation. It can provide several benefits such as voltage profile improvement, line loss reduction, and power factor correction. We know capacitor provides the reactive power, but its current has both active and reactive components with respect to the system reference. Capacitive reactance can cause resonance with the system inductance resulting in high harmonic voltage or current depending on the parallel or series resonance.

In this paper, in Section 2, it was analyzed how downstream node voltage can be higher than the voltage at upstream node. Section 3 demonstrates how negative line losses appear in a single phase in classical approach of line loss calculation. In Section 4, it was presented that capacitor current has both active and reactive components with respect to the system refer-ence, even though a capacitor generates only reactive power. In Section 5, the impact of harmonic resonance on the capacitor, series and parallel resonances, and harmonic mitigation technique has been discussed.

2. Analogy on higher voltage at downstream node

For a DC system, current flows from high voltage to low voltage, but that does not apply to an AC system shown in **Figure 1**, because in the study, it was seen that even though current flows from upstream node to downstream node, voltage at downstream node is higher than the voltage at upstream node. The reason of having higher voltage at downstream node in an AC system is that voltage drop in a single phase depends on mutual impedance and current in other phases besides self-impedance and current of its own phase as shown in Eq. (1).

$$
\begin{bmatrix} V_{agm} \\ V_{bgm} \\ V_{cgm} \end{bmatrix} = \begin{bmatrix} V_{agn} \\ V_{bgn} \\ V_{cgn} \end{bmatrix} - \begin{bmatrix} Z_{aa} & Z_{ab} & Z_{ac} \\ Z_{ba} & Z_{bb} & Z_{bc} \\ Z_{ca} & Z_{cb} & Z_{cc} \end{bmatrix} \begin{bmatrix} I_a \\ I_b \\ I_c \end{bmatrix}
\tag{1}
$$

$$
V_{igm} = V_{ign} - \Delta V_i
\tag{2}
$$

Now,

$$
|V_{ign}| = \sqrt{\left(\Re e[V_{ign}]\right)^2 + \left(\operatorname{Im}g[V_{ign}]\right)^2}
\tag{3}
$$

$$
|V_{igm}| = \sqrt{\left(\Re e[V_{ign}] - \Re e[\Delta V_i]\right)^2 + \left(\operatorname{Im}g[V_{ign}] - \operatorname{Im}g[\Delta V_i]\right)^2}
\tag{4}
$$

where V_{ign} and V_{igm} are the voltages of phase $i (i = a, b, c)$ at the upstream node and downstream node, respectively; ΔV_i is voltage drop in phase i; $\Re e[V_{ign}]$ and $\operatorname{Im}g[V_{ign}]$ are, respectively, real and imaginary components of V_{ign}; $\Re e[\Delta V_i]$ and $\operatorname{Im}g[\Delta V_i]$ are, respectively, real and imaginary components of ΔV_i.

Phase voltage at the downstream will be higher than the phase voltage at the upstream node if

$$
|V_{igm}| > |V_{ign}|
$$
$$
\Rightarrow \sqrt{\left(\Re e[V_{ign}] - \Re e[\Delta V_i]\right)^2 + \left(\operatorname{Im}g[V_{ign}] - \operatorname{Im}g[\Delta V_i]\right)^2} > \sqrt{\left(\Re e[V_{ign}]\right)^2 + \left(\operatorname{Im}g[V_{ign}]\right)^2}
$$
$$
\Rightarrow \left(\Re e[\Delta V_i]\right)^2 + \left(\operatorname{Im}g\Delta V_i\right)^2 - 2^* \Re e[V_{ign}]^* \Re e[\Delta V_i] - 2^* \operatorname{Im}g[V_{ign}]^* \operatorname{Im}g[\Delta V_i] > 0
\tag{5}
$$

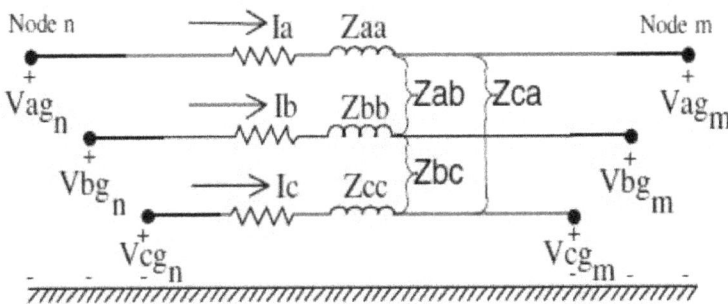

Figure 1. Three-phase line.

Similarly, phase voltage at downstream will be lower than the voltage at upstream node if

$$|V_{igm}| < |V_{ign}|$$

$$\sqrt{\left(\mathfrak{Re}[V_{ign}] - \mathfrak{Re}[\Delta V_i]\right)^2 + \left(\operatorname{Im} g[V_{ign}] - \operatorname{Im} g[\Delta V_i]\right)^2} < \sqrt{\left(\mathfrak{Re}[V_{ign}]\right)^2 + \left(\operatorname{Im} g[V_{ign}]\right)^2}$$

$$\Rightarrow (\mathfrak{Re}[\Delta V_i])^2 + (\operatorname{Im} g\Delta V_i)^2 - 2^* \mathfrak{Re}[V_{ign}] * \mathfrak{Re}[\Delta V_i] - 2^* \operatorname{Im} g[V_{ign}] * \operatorname{Im} g[\Delta V_i] < 0$$

$$(6)$$

Now, in Eq. (4), if real components of upstream node voltage (V_{ign}) and voltage drop (ΔV_i) have the same sign and imaginary components of upstream node voltage (V_{ign}) and voltage drop (ΔV_i) also have the same sign, then upstream node voltage must be higher than downstream node voltage. Similarly, in Eq. (4), if real components of upstream node voltage (V_{ign}) and voltage drop (ΔV_i) have opposite sign and imaginary components of upstream node voltage (V_{ign}) and voltage drop (ΔV_i) also have opposite sign, then upstream node voltage must be lower than downstream node voltage. Again, in Eq. (4), if real components of upstream node voltage (V_{ign}) and voltage drop (ΔV_i) have the same sign, but imaginary components of upstream node voltage (V_{ign}) and voltage drop (ΔV_i) have the opposite sign or if real components of upstream node voltage (V_{ign}) and voltage drop (ΔV_i) have the opposite sign, but imaginary components of upstream node voltage Vig_n and voltage drop (ΔV_i) have the same sign, then the upstream node voltage can be higher or lower than the downstream bus voltage. **Figure 2** shows how (+ and −) signs of real and imaginary components of upstream node voltage (V_{ign}) and voltage drop (ΔV_i) change for their locations in four quadrants.

If upstream node's voltage phasor (V_{ign}) and voltage drop phasor (ΔV_i) lie in the same quadrant, then real components of upstream node voltage (V_{ign}) and voltage drop (ΔV_i) will have the same sign, and imaginary components of V_{ign} and ΔV_i will also have the same sign; in this case, upstream node voltage must be higher than downstream node voltage.

If upstream node's voltage phasor (V_{ign}) and voltage drop phasor (ΔV_i) lie in two different quadrants which are exactly opposite (first and third quadrants or second and fourth quadrants), then real components of voltage (V_{ign}) and voltage drop (ΔV_i) will have opposite sign, and imaginary components of V_{ign} and ΔV_i will also have opposite sign; in this case, voltage at the

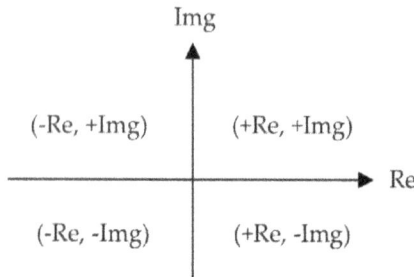

Figure 2. (+ and −) signs in four quadrants.

upstream node must be lower than the voltage at the downstream node. Again, if upstream node's voltage phasor (V_{ign}) and voltage drop phasor (ΔV_i) lie in two different quadrants which are adjacent (first and second quadrants or second and third quadrants or third and fourth quadrants or fourth and first quadrants), then real components of V_{ign} and ΔV_i can have the same sign or opposite sign; if they have the same sign, then imaginary components of V_{ign} and ΔV_i will have the opposite sign; if they have opposite sign, then imaginary components of V_{ign} and ΔV_i will have the same sign; in these both cases, upstream node voltage can be higher or lower than the downstream node voltage. **Table 1** summarizes the conditions that cause the upstream node voltage to be higher or lower than the downstream node voltage.

How the downstream node voltage and voltage drop vary with loads in an AC system is tested on a two-bus system in **Figure 3**, and results are shown in **Table 2**.

In **Table 2**, the phasor diagrams show how voltage drop and downstream node voltage change their locations into the quadrants as load varies. For example, when a three-phase load at node 2 changes from 1000 kW+ j500 kVAr, 500 kW+ j200 kVAr, and 300 kW+ j100 kVAr to 200 kW+ j100 kVAr, 500 kW+ j200 kVAr, and 300 kW+ j100 kVAr, respectively, in phase a, phase b, and phase c, it is seen that voltage drop phasor ΔV_a changes its location from first quadrant to fourth quadrant and voltage drop phasor ΔV_c changes its location from first quadrant to third quadrant.

When a three-phase load, 600 kW + j300kVAr, 500 kW + j200kVAr, and 300 kW + j100kVAr, is connected at the downstream node 2 in phase a, phase b, and phase c, respectively, it is seen that phase c voltage (2394.8 V) at the downstream node 2 is lower than the phase c voltage (2400 V) at the upstream node 1, where voltage phasor V_{cgn} at the upstream node 1 and voltage drop phasor ΔV_c lie in the same quadrant (second). Again, when the three-phase load changes

Location of upstream node voltage (Vig$_n$) and voltage drop (ΔVi) phasors into the quadrants		Upstream node voltage level with respect to downstream node voltage level
Same quadrant		$\|V_{ign}\| > \|V_{igm}\|$
Different quadrants	Exactly opposite quadrants	$\|V_{ign}\| < \|V_{igm}\|$
	Adjacent quadrants	$\|V_{ign}\| > \|V_{igm}\|$ or $\|V_{ign}\| < \|V_{igm}\|$

Table 1. Conditions for the upstream node voltage to be higher or lower than downstream node voltage.

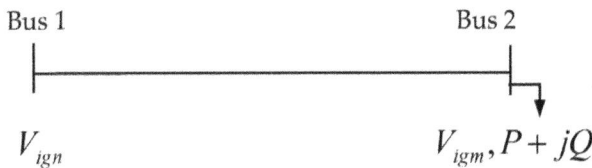

Figure 3. Two-bus system.

	Bus voltage (V)	Phasor diagram of bus voltages and voltage drop	Voltage drop (V)	Load (kW, kVAr)
Bus 1	2400∠0 (2400 + j0) 2400∠120 (−1200 − j 2078.5) 2400∠120 (−1200 + j2078.5)			
Bus 2	2261.0∠−2.4 (2259 − j 95.4) 2426.4∠−121.8 (−1280.2 − j2061.2) 2386.9∠120.7 (−1218.8 + j2052.2)		140.98 + j95.44(170.2∠34.09) 80.22−j17.26(82.1∠−12.14) 18.75 + j26.22 (32.23∠54.4)	1000 + j500 500 + j200 300 + j100
	2292.4∠−1.7148 (2291.4 −j68.6) 2406.7∠−121.7 (−1265.1 −j2047.3) 2389.1∠120.28 (−1205 + j2063)		108.59 + j68.61 (128.4∠32.3) 65.12−j31.13 (72.2∠−25.55) 4.97 + j15.5 (16.27∠72.22)	800 + j400 500 + j200 300 + j100
	2351.7∠−1.03 (2322.1−j41.6) 2368.1∠−121.6 (−1250.9 −j2033.4) 2394.8∠119.9 (−1191.9 + j2073.8)		77.89 + j 41.56 (88.3∠28.08) 50.86−j 45.1 (67.97∠−41.56) −8.103 + j 4.64 (9.3∠150.17)	600 + j300 500 + j200 300 + j100
	2351.7∠−0.358 (2351.6−j14.7) 2368.1∠−121.48 (−1236.9 −j2019.5) 2394.8∠119.5 (−1179.3 + j2084.3)		48.37 + j 14.7 (50.5∠16.89) 36.88−j 58.9 (69.5∠−57.98) −20.67 −j 5.84 (21.5∠−164.2)	400 + j200 500 + j200 300 + j100

Bus voltage (V)	Voltage drop (V)	Phasor diagram of bus voltages and voltage drop	Load (kW, kVAr)
2365.90∠−0.029 (2365.9−j1.2) 2358.7∠−121.4 (−1230.3−j2012.5) 2396.7∠119.3 (−1173.2 + j2089.9)	34.13 + j 1.15 (34.15∠1.94) 30.284−j 65.99 (∠−65.35) −26.78−j11.4 (72.6∠−156.8)		300 + j150 500 + j200 300 + j100
2379.9∠0.29 (2379.9 + j12.1) 2349.6∠−121.4 (−1223.8−j2005.7) 2398.3∠119.1 (−1167.3 + j2095.0)	20.14−j 12.1 (23.5∠−30.9) 23.84−j 72.78 (76.5∠−71.8) −32.7−j 16.58 (36.6∠−153.1)		200 + j100 500 + j200 300 + j100
2394.0∠0.62 (2393.8 + j25.9) 2339.6∠−121.33 (−1216.7−j1998.3) 2400.4∠118.9 (−1161.2 + j2100.9)	6.15−j25.86 (26.58∠−76.6) 16.7−j80.12 (81.8∠−78.2) −38.8−j22.4 (44.8∠−150.02)		100 + j50 500 + j200 300 + j100
2400.1∠0.776 (2399.9 + j 0.032.5) 2335.8∠−121.3 (−1214.6−j1995.2) 2401.3∠118.8 (−1158.8 + j2103.2)	0.1−j32.52 (32.52∠−89.8) 14.6−j83.26 (84.5∠−80.04) −41.2−j24.76 (48.06∠−149)		50 + j25 500 + j200 300 + j100

Bus 1 phase a voltage: ⟶ Bus 2 phase a voltage: -----⟶ Voltage drop in phase a: ⟶
Bus 1 phase b voltage: ·······⟶ Bus 2 phase b voltage: ·······⟶ Voltage drop in phase b: ·······⟶
Bus 1 phase c voltage: ⟶ Bus 2 phase c voltage: ---- ⟶ Voltage drop in phase c: ·····⟶

Table 2. Variation in the location of node voltage phasors and voltage drop phasors with loads.

to 100 kW + j50kVAr, 500 kW + j200kVAr, and 300 kW + j100kVAr in respective phases, it is seen that phase c voltage (2400.4 V) at the downstream node 2 is higher than phase c voltage (2400 V) at the upstream node 1, where voltage phasor V_{cgn} at the upstream node 1 lies in second quadrant but voltage drop phasor ΔV_c lies in third quadrant.

3. Analogy on negative line loss

A classical approach of line loss calculation was presented here. In this approach, line loss was computed as the difference between input power at upstream node and output power at downstream node. The formulation of line loss equation and condition for negative line loss has been demonstrated using a two-bus system shown in **Figure 4**.

Line loss = input power – output power

$$
\begin{aligned}
P_{Loss} + jQ_{Loss} &= V_n I^* + V_m I^* \\
&= |V_n|\angle\alpha_n{}^*|I|\angle - \beta - |V_m|\angle\alpha_m{}^*|I|\angle - \beta \\
&= |V_n||I|\angle(\alpha_n - \beta) - |V_m||I|\angle(\alpha_m - \beta) \\
&= |V_n||I|\angle\theta_n - |V_m||I|\angle\theta_m
\end{aligned}
\tag{7}
$$

Active power loss,

$$
P_{Loss} = |V_n||I|\cos\theta_n - |V_m||I|\cos\theta_m
\tag{8}
$$

where input power is the power leaving the upstream node and output power is the power entering into the downstream node. $|V_n|\angle\alpha_n$ is the voltage at the upstream node n; $|V_m|\angle\alpha_m$ is the voltage at the downstream node m; $|I|\angle\beta$ is the current in the line between the node n and m; and $\theta_n = (\alpha_n - \beta)$ and $\theta_m = (\alpha_m - \beta)$ are the power factor angles at node n and m, respectively.

From Eq. (8), we can see that active line loss will be negative, if

$$
\begin{aligned}
&P_{Loss} < 0 \\
\Rightarrow\ &if, |V_n||I|\cos\theta_n - |V_m||I|\cos\theta_m < 0 \\
\Rightarrow\ &if, |V_n|\cos\theta_n < |V_m|\cos\theta_m
\end{aligned}
\tag{9}
$$

Here, four cases are considered to demonstrate the conditions of occurrence of negative line losses:

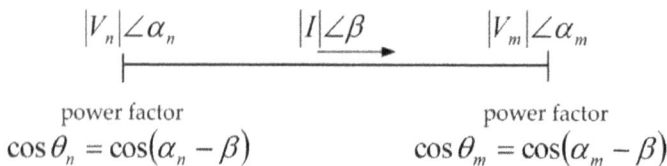

$$
|V_n|\angle\alpha_n \qquad\qquad |I|\angle\beta \qquad\qquad |V_m|\angle\alpha_m
$$

power factor
$$\cos\theta_n = \cos(\alpha_n - \beta)$$

power factor
$$\cos\theta_m = \cos(\alpha_m - \beta)$$

Figure 4. A line segment between upstream node n and downstream node m.

Case 1: Downstream node voltage lags behind the upstream node voltage and leads the line current in **Figure 5**.

As voltage $|V_n|\angle\alpha_n$ leads voltage $|V_m|\angle\alpha_m$, we have

$$\alpha_n > \alpha_m$$

$$\Rightarrow (\alpha_n - \beta) > (\alpha_m - \beta)$$

$$\Rightarrow \cos(\alpha_n - \beta) < \cos(\alpha_m - \beta)$$

$$\Rightarrow \cos(\theta_n) < \cos(\theta_m) \tag{10}$$

Since by Eq. (10), $\cos\theta_n < \cos\theta_m$ in Eq. (9), there is a possibility of $|V_n|\cos\theta_n < |V_m|\cos\theta_m$, i.e., $P_{Loss} < 0$, even though $|V_n| > |V_m|$. If upstream node voltage is lower than the downstream node voltage ($|V_n| < |V_m|$) that may happen in an AC system as shown in Section 2, then it confirms that $|V_n|\cos\theta_n < |V_m|\cos\theta_m$ (i.e., active line loss) must be negative.

Case 2: Downstream node voltage leads both the upstream node voltage and the line current as shown in **Figure 6**.

As voltage $|V_n|\angle\alpha_n$ lags the voltage $|V_m|\angle\alpha_m$, we have

$$\alpha_n < \alpha_m$$

$$\Rightarrow (\alpha_n - \beta) < (\alpha_m - \beta)$$

$$\Rightarrow \cos(\alpha_n - \beta) > \cos(\alpha_m - \beta)$$

$$\Rightarrow \cos(\theta_n) > \cos(\theta_m) \tag{11}$$

Even though, by Eq. (11), $\cos\theta_n > \cos\theta_m$ in Eq. (9), there is still a possibility of $|V_n|\cos\theta_n < |V_m|\cos\theta_m$ (i.e., $P_{Loss} < 0$) if upstream node voltage is lower than the downstream node voltage $|V_n| < |V_m|$ which could happen very rarely. If the upstream node voltage is higher than the

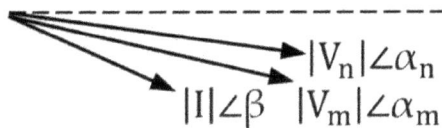

Figure 5. Phasor diagram for upstream and downstream node voltage and line current for Case 1.

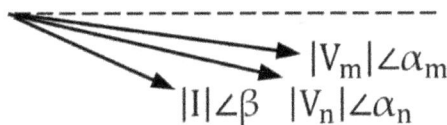

Figure 6. Phasor diagram for upstream and downstream node voltage and line current for Case 2.

downstream node voltage, $|V_n| > |V_m|$, then it is guaranteed that active line loss cannot be negative.

Case 3: Downstream node voltage lags behind both the upstream node voltage and the line current, as shown in **Figure 7**.

As voltage $|V_n|\angle\alpha_n$ leads voltage $|V_m|\angle\alpha_m$, we have

$$\alpha_n > \alpha_m$$

$$\Rightarrow (\alpha_n - \beta) > (\alpha_m - \beta)$$

[Since $(\alpha_n - \beta) < 0$ and $(\alpha_m - \beta) < 0$, multiply both sides by -1]

$$\Rightarrow -(\alpha_n - \beta) < -(\alpha_m - \beta)$$

$$\Rightarrow \cos\left(-(\alpha_n - \beta)\right) > \cos\left(-(\alpha_m - {}^m\beta)\right)$$

[Since $\cos\left(-(\alpha_n - \beta)\right) = \cos(\alpha_n - \beta)$ and $\cos\left(-(\alpha_m - \beta)\right) = \cos(\alpha_m - \beta)$

$$\Rightarrow \cos(\theta_n - \beta) > \cos(\theta_m - \beta)$$

$$\Rightarrow \cos(\theta_n) > \cos(\theta_m) \tag{12}$$

Even though, by Eq. (12), $\cos\theta_n > \cos\theta_m$ in Eq. (9), there is still a chance of $|V_n|$ $\cos\theta_n < |V_m|\cos\theta_m$, i.e., $P_{Loss} < 0$, if downstream node voltage is higher than the upstream node voltage ($|V_n| < |V_m|$). If downstream node voltage is lower than the upstream node voltage, ($|V_n| > |V_m|$), then it confirms that active line loss cannot be negative.

Case 4: Downstream node voltage leads the upstream voltage and lags behind the line current, as shown in **Figure 8**.

As voltage $|V_n|\angle\alpha_n$ lags voltage $|V_m|\angle\alpha_m$, we have

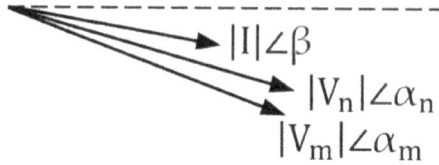

Figure 7. Phasor diagram for upstream and downstream node voltage and line current for Case 3.

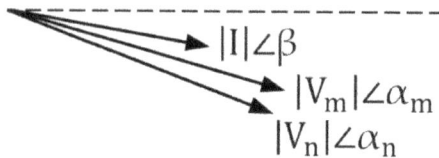

Figure 8. Phasor diagram for upstream and downstream node voltage and line current for Case 4.

$$\alpha_n < \alpha_m$$

$$\Rightarrow (\alpha_n - \beta) < (\alpha_m - \beta)$$

[Since $(\alpha_n - \beta) < 0$ and $(\alpha_m - \beta) < 0$, multiply both sides by -1]

$$\Rightarrow -(\alpha_n - \beta) > -(\alpha_m - \beta)$$

$$\Rightarrow \cos\left(-(\alpha_n - \beta)\right) < \cos\left(-(\alpha_m - {}^m\beta)\right)$$

[Since $\cos\left(-(\alpha_n - \beta)\right) = \cos(\alpha_n - \beta)$ and $\cos\left(-(\alpha_m - \beta)\right) = \cos(\alpha_m - \beta)$]

$$\Rightarrow \cos(\theta_n - \beta) < \cos(\theta_m - \beta)$$

$$\Rightarrow \cos(\theta_n) < \cos(\theta_m) \tag{13}$$

Since, by Eq. (13), $\cos\theta_n < \cos\theta_m$ in Eq. (9), there is a possibility of $|V_n|\cos\theta_n < |V_m|\cos\theta_m$, i.e., $P_{Loss} < 0$, even though $(|V_n| > |V_m|)$. If downstream node voltage is higher than the upstream node voltage $|V_n| < |V_m|$, then it is confirmed that $|V_n|\cos\theta_n < |V_m|\cos\theta_m$, i.e., active line loss must be negative.

By and large, downstream node's power factor will be greater than upstream node's power factor if downstream node voltage lags behind the upstream node voltage and leads the line current or downstream node voltage leads the upstream voltage and lags behind the line current. In this case, if upstream node voltage is lower than downstream node voltage, then output power at downstream node will be higher than input power at upstream node, i.e., line loss must be negative.

Downstream node's power factor will be lower than upstream node's power factor, if downstream node voltage leads both the upstream node voltage and the line current or downstream node voltage lags behind both the upstream node voltage and the line current. In this case, there is a chance of getting a negative line loss, only if upstream node voltage is lower than downstream node voltage. **Table 3** summarizes the chances of occurrence of negative line loss in the classical approach-based line loss calculation.

Generally, transmission and distribution lines are inductive by nature; therefore, downstream node voltage leads the line current and lags behind the upstream node voltage. Downstream node voltage may lead the upstream node voltage in a lightly loaded phase in an unbalanced system. Both upstream and downstream node voltages can lag behind the line current for a system with too many capacitor banks for voltage profile improvements and/or line loss reduction.

Example 1: IEEE's load flow results in [6] show that, for 13-bus system shown in **Figure 9**, line loss in phase b feeder between substation voltage regulator's secondary side and node 632 is negative (-3.25 kW), which can be explained by Eq. (9). From IEEE's load flow results, phase b voltage at substation voltage regulator's secondary is $V_{bRG60} = 2521.86\angle - 120$; phase b voltage at node 632 is $V_{b632} = 2502.65\angle - 121.72$; and phase b current in that feeder is $I = 414.37\angle - 140.91$.

Phasor diagram of V_{bRG60}, V_{b632}, and I is shown in **Figure 10**.

Lagging/leading status	Power factor (Pf)	Voltage level status	Occurrence of negative line loss
Downstream node voltage lags the upstream node voltage and leads the line current or downstream node voltage leads the upstream node voltage and lags the line current	Pf at downstream node > Pf at upstream node	If downstream node voltage < upstream node voltage	Possibility of occurrence of negative line loss
		If downstream node voltage > upstream node voltage	Guarantee of occurrence of negative line loss
Downstream node voltage leads both the upstream node voltage and the line current or downstream node voltage lags behind both the upstream node voltage and the line current	Pf at downstream node < Pf at upstream node	If downstream node voltage < upstream node voltage	No possibility of occurrence of negative line loss
		If downstream node voltage > upstream node voltage	Possibility of occurrence of negative line loss

Table 3. Conditions of occurrence of negative line loss.

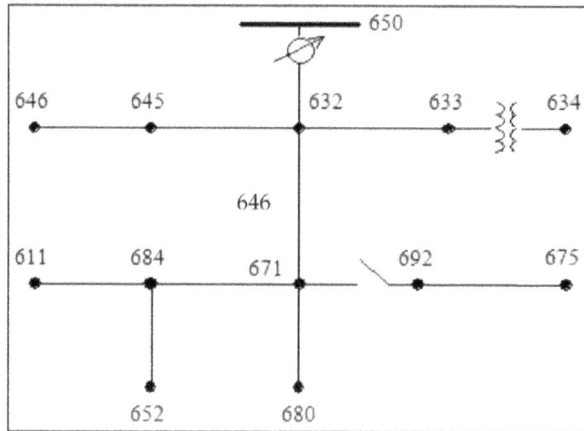

Figure 9. IEEE 13-bus system.

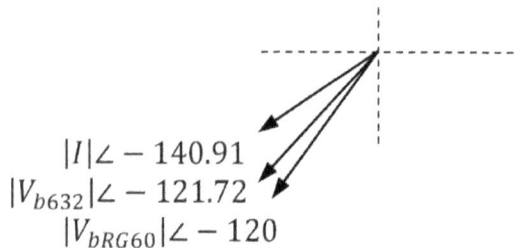

$$|I| \angle -140.91$$
$$|V_{b632}| \angle -121.72$$
$$|V_{bRG60}| \angle -120$$

Figure 10. Phasor diagram of voltages and line current for the feeder between substation regulator and node 632.

By Eq. (8), active power loss is

$$P_{Loss} = |V_{bRG60}||I| \cos \theta_{bRG60} - |V_{b632}||I| \cos \theta_{b632}$$

$$= 2521.86*414.37*\cos(20.91) - 2502.65*414.37*\cos(19.19)$$

$$= 2521.86*414.37*0.93414 - 2502.65*414.37*0.9444$$

$$= 1044.985*0.93414 - 1037.023*0.9444 \text{ kW}$$
$$\downarrow \qquad\qquad\qquad \downarrow$$

The upstream node's power factor is less than the downstream node's power factor

$$= 976.16 - 979.36 \text{ kW}$$

$$= -3.2 \text{ kW}$$

The above power loss calculation shows that, although downstream node voltage (2502.65 V) is lower than upstream node voltage (2521.86 V), downstream node's power factor (0.9444) is higher than upstream node's power factor (0.93414). The upstream node's lower power factor makes the input power (976.16 kW = 2521.86*414.37*0.93414) lower than the output power (979.36 kW = 2502.65*414.37*0.9444) at the downstream node, even though all the active power at the downstream node comes from the upstream node. The reason of higher power factor at the downstream node is that downstream node lags behind the upstream node voltage and leads the line current as shown in **Figure 10**.

Example 2: IEEE's load flow results in [6] also show that, for 123-bus system shown in **Figure 11**, line loss in phase b feeder between node 101 and node 105 is negative (−0.019 kW), which can also be explained by Eq. (9). From IEEE's load flow results, phase b voltage at node

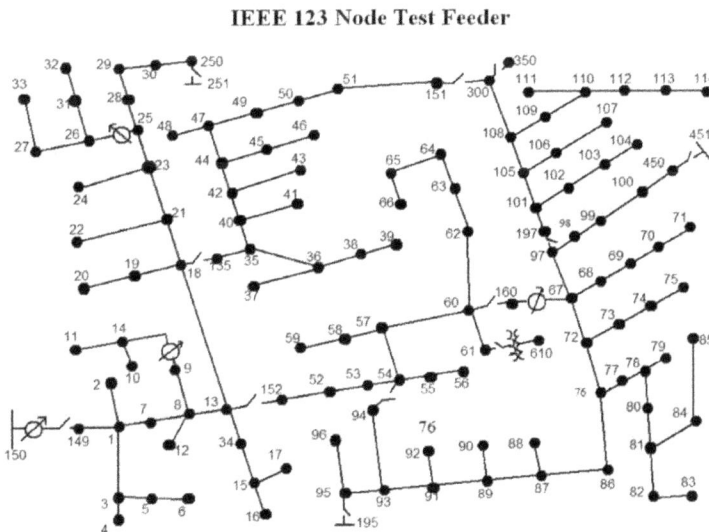

Figure 11. IEEE 123-bus system.

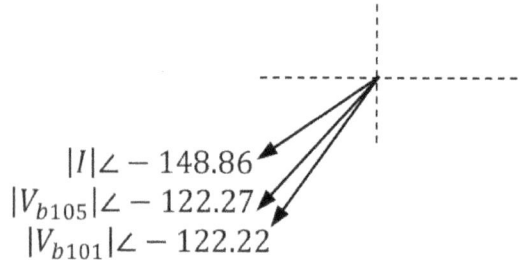

$$|I|\angle - 148.86$$
$$|V_{b105}|\angle - 122.27$$
$$|V_{b101}|\angle - 122.22$$

Figure 12. Phasor diagram of voltages and line current for the line between node 101 and node 105.

101 is $V_{b101} = 2474.55\angle - 122.22V_{b101} = 2474.55\angle - 122.22$; phase b voltage at node 105 is $V_{b105} = 2474.07\angle - 122.27V_{b105} = 2474.07\angle - 122.27$; and phase b current in that feeder is $I = 36.21 \angle - 148.86I = 36.21\angle - 148.86$.

Phasor diagram of V_{b101}, V_{b105}, and I is shown in **Figure 12**.

By Eq. (8), active power loss is

$$P_{Loss} = |V_{b101}||I| \cos\theta_{b101} - |V_{b105}||I| \cos\theta_{b105}$$

$$= 2474.55*36.21*\cos(26.64) - 2474.07*36.21*\cos(26.59)$$

$$= 2474.55*36.21*0.893841424 - 2474.07*36.21*0.894232372$$

$$= 89603.455*0.893841424 - 89586.0747*0.894232372 \text{ kW}$$
$$\downarrow \qquad\qquad\qquad\qquad \downarrow$$

The upstream node's power factor is less than the downstream node's power factor

$$= 80.09128026 - 80.11076808 \text{ kW}$$

$$= -0.019487 \text{ kW}$$

In the line loss calculation above, it is seen that, although voltage (2474.55 V) at the upstream node is higher than the voltage (2474.07 V) at the downstream node, power factor (0.893841424) at the upstream node is less than the power factor (0.894232372) at the downstream node. The higher power factor at the downstream node makes the output power (80.11076808 kW = $2474.07*36.21*0.894232372*10^{-3}$) at the downstream node greater than the input power (80.09128026 kW = $2474.55*36.21*0.893841424*10^{-3}$) at the upstream node, even though all the active power at the downstream node comes from the upstream node. Greater output power at downstream node with respect to input power at upstream node is the reason of having a negative line loss.

4. Active and reactive components of capacitor current

Although a capacitor delivers reactive power, its current has both active and reactive components with respect to the system reference. The reason of having both active and reactive current components with respect to the system reference is that capacitor current leads the capacitor bus voltage by 90°, with bus voltage angle represented with respect to the system reference. Now the angle of a capacitor current with respect to the system reference is equal to the capacitor bus voltage angle plus 90°; the cosine and sine value of that angle of capacitor current make, respectively, the active and reactive current component of the capacitor current, where capacitor's current is only reactive with respect to capacitor bus voltage. Here it needs to be mentioned that if Kirchhoff's current law is applied at the capacitor bus, considering capacitor current is only reactive, then it does not satisfy the Kirchhoff's current law. If the angle of capacitor current is presented with respect to the system reference which results in both real and imaginary components of capacitor current, then it satisfies Kirchhoff's current law. These have been demonstrated with the load flow results of 13-bus system. Load flow results [6] of 13-bus system are given below for phase a of node 675 and line 692–675:

Voltage at node 675 is $V_{675} = 0.9835\angle{-5.56°}$; current in line 692–675 is $I_{Line} = 205.33\angle{-5.15°}$; load that is constant power load at node 675 is $P_{Load} + jQ_{Load} = 485 + j190$; and output power of the capacitor is $jQ_{Load} = j193.4$.

Now, line current: $I_{Line} = 205.33\angle{-5.15°} = 204.501 - j18.43$

Load current: $|I_{Load}|^2 = \frac{P_{Load}^2 + Q_{Load}^2}{|V|^2}$

$$= \frac{485^2 + 190^2}{|0.9835*2.40177|^2} = 48626.8$$

$$\Rightarrow |I_{Load}| = 205.33$$

If θ is the power factor angle, i.e., angle between voltage and current, then

$$\cos\theta = \frac{485}{\sqrt{485^2 + 190^2}} = 0.931,$$

$$\Rightarrow \theta = 21.39$$

Since the load current lags the voltage, we have

$$I_{Load} = 220.5148\angle(-5.56° - 21.39°) = 196.56 - j99.95$$

Capacitor current: $I_{Cap} = \frac{Q_{Cap}}{|V|}$

$$= \frac{193.44}{0.9835*2.40177} = 81.8915$$

Here, if the capacitor current is assumed to be only reactive current, i.e., $I_{Cap} = 81.89\angle{90°}$, then it does not satisfy Kirchhoff's current law ($\sum linecurrent + loadcurrent + capacitorcurrent = 0$) at the bus 675. Capacitor current I_{Cap} leads bus voltage V_{675} ($0.9835\angle{-5.56°}$) by 90°; therefore, the angle between the system reference and capacitor current I_{Cap} is $-5.56° + 90° = 84.44°$, and I_{Cap} should be presented as $I_{Cap} = 81.89\angle{84.44°} = 7.93 + j81.51$. It is seen that capacitor current I_{Cap} does not make a 90° angle with the system reference which is the phase a voltage of substation regulator. Therefore, capacitor current has both real and imaginary current components with

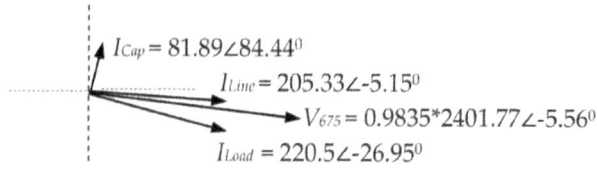

Figure 13. Phasor diagram of bus voltage, line current, and capacitor current.

respect to the system reference, whereas the output power of a capacitor is reactive. It needs to be mentioned that capacitor current is also only reactive with respect to capacitor bus voltage. **Figure 13** depicts the phasor diagram of I_{Line}, I_{Load}, I_{Cap}, and V_{675}.

Kirchhoff's current law is satisfied with the capacitor current, I_{Cap} = 81.89∠84.44°. It is seen at bus 675 in phase a:

$$P_{Line} + jQ_{Line} = V_{675}I^*_{Line}$$

$$= 0.9835^*2.4017\angle - 5.56^*205.3\angle - 5.15$$
$$= 485 + j3.47$$

If we apply the law of conservation of power at node 675 in phase a, we can see

$$\left(P_{Line} + jQ_{Line}\right) + \left(jQ_{cap}\right) = \left(\underline{\underline{P_{Load} + jQ_{Load}}}\right)$$
$$\Rightarrow (485 - j3.4) + (j193.44) = (485 + j190)$$

Here we can see that power consumed by the load is equal to the capacitor's output power plus power injected through the line, whereas line current is equal to capacitor current plus load current.

5. Impact of harmonic resonance on capacitor

Shunt capacitors are used for voltage profile improvement and power factor correction. A capacitor can cause resonance frequency which results in high voltage across the capacitor and severe voltage distortion. Therefore, it needs to verify if the shunt capacitive reactance will resonate with the system inductive reactance. A harmonic resonance will occur if

$$X_L h = X_C h$$

$$2\pi fhL = \frac{1}{2\pi fhC}$$

$$fh = \frac{1}{2\pi\sqrt{LC}}$$

where X_L is the system reactance at fundamental frequency, X_C is the capacitive reactance at fundamental frequency, h is the harmonic order, and f is the fundamental frequency. L is the system inductance, and C is the capacitance of the shunt capacitor.

Formulation of relationship among harmonic order, capacitor-rated MVAR, and system fault MVA at the capacitor location:

At resonant condition,

$$2\pi fhL = \frac{1}{2\pi fhC}$$

$$h^2 = \frac{1}{2\pi fL} * \frac{1}{2\pi fC}$$

$$= \frac{1}{X_L} * X_C$$

$$= \frac{V*V*10^6}{X_L} * \frac{X_C}{V*V*10^6}$$

$$= I_{sc}*V*10^6 * \frac{1}{I_C*V*10^6}$$

$$= \frac{MVA_{SC}}{MVAR_C}$$

$$h = \frac{\sqrt{MVA_{SC}}}{\sqrt{MVAR_C}}$$

where V is the system voltage, I_{sc} is the available short circuit current at the capacitor location, and I_c is the rated current of the capacitor. MVA_{SC} is the available fault capacity at the capacitor location, and $MVAR_C$ is the rated reactive power of capacitor.

Harmonic current can have an adverse effect on the capacitor resulting in overloading, overheating, and voltage stress. As per IEEE 18-2000, capacitor shall deliver a maximum of 135% of its rated reactive power (kVAR). It also withstands a maximum continuous RMS overvoltage of 110%, peak overvoltage of 120%, and an overcurrent of 180% of rated value.

5.1. Series and parallel resonance

Whether a series or parallel resonance will occur or not depends on the system configuration. Nowadays the distribution system is a radial system. Motor loads in the system make the equivalent impedance lower at the capacitor location because at the short circuit, motor loads inject the current back to the system adding parallel circuits. Equivalent impedance of the system with dominant non-motor load is higher than the system with dominant motor load. Series and parallel resonance circuits are shown below (**Figures 14** and **15**).

5.1.1. Series resonance

$$I\angle\Theta = \frac{V\angle 0}{R + j2\pi fL - \frac{j}{2\pi fC}}$$

$$|I| = \frac{|V|}{\sqrt{R^2 + \left(2\pi fL - \frac{1}{2\pi fC}\right)^2}}, \angle\Theta = -\tan^{-1}\left(\frac{2\pi fL - \frac{1}{2\pi fC}}{R}\right)$$

$|I|$ and $\angle\Theta$ vary with f. When $2\pi fL = \frac{1}{2\pi fC}$, $|I|$ is maximum and $\angle\Theta = 0$, i.e., current and voltage are in phase.

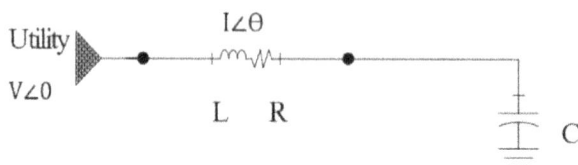

Figure 14. Series resonance circuit.

5.1.2. Parallel resonance

$$V\angle\Theta = \frac{I\angle 0}{Y}$$

where

$$Y = \frac{1}{R} + j2\pi fC + \frac{1}{j2\pi fL}$$

$$|V| = \frac{|I|}{\sqrt{\left(\frac{1}{R}\right)^2 + \left(2\pi fC - \frac{1}{2\pi fL}\right)^2}}, \angle\Theta = -\tan^{-1}\left(\frac{2\pi fC - \frac{1}{2\pi fL}}{\frac{1}{R}}\right)$$

$|V|$ and $\angle\Theta$ vary with f. When $2\pi fC = \frac{1}{2\pi fL}$, $|V|$ is maximum and $\angle\Theta = 0$, i.e., current and voltage are in the same phase.

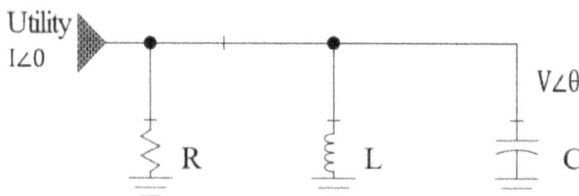

Figure 15. Series resonance circuit.

5.2. Harmonic mitigation methods

There are several techniques to mitigate harmonic distortions. The following are the prevailing methods for mitigating the harmonic distortions:

(i) Reactor: Reactor is a simple and cost-effective technique to reduce the harmonics injected by nonlinear loads. Reactors are usually used to the nonlinear loads such as variable speed drives. The changing current through a reactor induces voltage across its terminals in the opposite direction of the applied voltage which consequently opposes the rate of change of current. This characteristic of a reactor helps in reducing the harmonic currents produced by variable speed drives and other nonlinear loads.

(ii) Delta–delta and delta-wye transformers: In this technique, two separate utility trans-formers with equal nonlinear loads are used. The phase relationship to various six-pulse converters is shifted through cancelation techniques that help in reducing the harmonics. This technique is also used in a 12-pulse front end of the drive.

(iii) Isolation transformers: In this technique, system voltage is stepped up or stepped down for voltage match. A neutral ground reference for nuisance ground faults is also provided in this technique. This is the best solution when SCRs are used as bridge rectifiers in AC and DC drive.

(iv) Passive harmonic filters or line harmonic filters: Passive or line harmonic filters (LHF) are used to mitigate lower order harmonics such as fifth, seventh, eleventh, and thirteenth. This also known as harmonic trap filters. In a six-pulse drive, it is used as a stand-alone part. It is also used for multiple single-phase nonlinear loads. In line harmonic filters, LCR circuit is tuned to a particular harmonic frequency that needs to be mitigated. Their operation is based on the resonance phenomena.

6. Conclusion

In this paper, it was presented how higher voltage at downstream node and negative line losses appear in an AC power system. It was also demonstrated that capacitor current has both active and reactive components with respect to the system reference.

If the upstream bus voltage and voltage drop phasors lie in the different quadrants that are exactly opposite to each other, then downstream node voltage will be higher than upstream bus voltage; if the upstream bus voltage and voltage drop phasors lie in the same quadrant, then upstream node voltage will be higher than downstream node voltage.

Transmission and distribution lines are generally inductive by nature; therefore, downstream node voltage leads the line current and lags behind the upstream node voltage, and that results in downstream node's power factor to be greater than the upstream node's power factor. Downstream node's higher power factor can cause the output power to be greater than the input power at upstream node, although all the power comes from the upstream node. If input power at upstream node is lower than output power at downstream node, then line loss will be negative in classical approach-based line loss calculation, where line losses are calculated as the difference between input power at upstream node and output power at downstream node. If the upstream node's power factor is higher than the downstream node's power factor and upstream node voltage is lower than downstream node voltage, then there is still a chance of having a negative line loss.

A capacitor supplies only reactive power, but its current has both real and imaginary components with respect to the system reference. With respect to the capacitor bus voltage, all the current of a capacitor is reactive. Therefore, real and imaginary current components are reference relative.

Shunt capacitor used for power factor correction and voltage profile improvement can cause resonance at a harmonic frequency with system inductive reactance. At series resonance capacitor, current is very high, and at parallel resonance, voltage across the capacitor is very high which can rupture the capacitor. Therefore, the size of the capacitor should be checked if resonant condition can occur. There are several methods to mitigate the harmonic distortion — the use of reactor, delta–delta and delta-wye transformer, isolation transformers, and passive harmonic filters or line harmonic filters. A reactor is the simple and cost-effective method to reduce the harmonic distortion.

Author details

Sushanta Paul

Address all correspondence to: sxpaul@shockers.wichita.edu

Wichita State University, Wichita, USA

References

[1] Vani E, Rengarajan N. Improving the power quality of the wind power system using low cost topology. International Journal of Modelling and Simulation. 2017;37(2):108-115

[2] Carpaneto E, Chicco G, Akilimali JS. Loss partitioning and loss allocation in three-phase radial distribution systems with distributed generation. IEEE Transactions on Power Systems. Aug. 2008;23(3):1039-1049

[3] Kersting WH. The computation of neutral and dirt currents and power losses. Proceedings of Conference on IEEE PES, Transmission and Distribution. Sept. 2003;3:978-983

[4] Paul SA, Jewell W. Impact of load type on power consumption and line loss in voltage reduction program. In: Proceedings of Conference on North American Power Symposium (NAPS). 2013. pp. 1-6

[5] Paul S, Jewell W. 24 Factorial design for joint effect of ambient temperature and capacitor price, size and, phase kVAr on line loss. Proceedings of Conference on IEEE PES General Meeting; July, 2014. pp. 1-5

[6] IEEE PES Distribution System Analysis Subcommittee's Distribution Test Feeder Working Group. URL: http://ewh.ieee.org/soc/pes/dsacom/testfeeders/index.html